Communications
in Computer and Information Science   47

José Cordeiro   Boris Shishkov
AlpeshKumar Ranchordas
Markus Helfert (Eds.)

Software and
Data Technologies

Third International Conference, ICSOFT 2008
Porto, Portugal, July 22-24, 2008
Revised Selected Papers

 Springer

Volume Editors

José Cordeiro
Polytechnic Institute of Setúbal and INSTICC
Rua do Vale de Chaves, Estefanilha, 2910-761, Setúbal, Portugal
E-mail: jcordeir@est.ips.pt

Boris Shishkov
IICREST
P.O. Box 104, 1618 Sofia, Bulgaria
E-mail: b.b.shishkov@tudelft.nl

AlpeshKumar Ranchordas
INSTICC
Avenida D. Manuel I, 2910-595 Setúbal, Portugal
E-mail: alpesh.ranchordas@gmail.com

Markus Helfert
Dublin City University, School of Computing, Dublin 9, Ireland
E-mail: Markus.Helfert@computing.dcu.ie

Library of Congress Control Number: 2009936865

CR Subject Classification (1998): D.2, D.4.6, K.6.5, H.2, K.8, E.3, E.2, I.3.7

ISSN   1865-0929
ISBN-10   3-642-05200-2 Springer Berlin Heidelberg New York
ISBN-13   978-3-642-05200-2 Springer Berlin Heidelberg New York

springer.com

© Springer-Verlag Berlin Heidelberg 2009
Printed in Germany

Typesetting: Camera-ready by author, data conversion by Scientific Publishing Services, Chennai, India
Printed on acid-free paper   SPIN: 12771264   06/3180   5 4 3 2 1 0

Preface

This book contains the best papers of the Third International Conference on Software and Data Technologies (ICSOFT 2008), held in Porto, Portugal, which was organized by the Institute for Systems and Technologies of Information, Communication and Control (*INSTICC*), co-sponsored by the Workflow Management Coalition (WfMC), in cooperation with the Interdisciplinary Institute for Collaboration and Research on Enterprise Systems and Technology (IICREST).

The purpose of ICSOFT 2008 was to bring together researchers, engineers and practitioners interested in information technology and software development. The conference tracks were "Software Engineering", "Information Systems and Data Management", "Programming Languages", "Distributed and Parallel Systems" and "Knowledge Engineering".

Being crucial for the development of information systems, software and data technologies encompass a large number of research topics and applications: from implementation-related issues to more abstract theoretical aspects of software engineering; from databases and data-warehouses to management information systems and knowledge-base systems; next to that, distributed systems, pervasive computing, data quality and other related topics are included in the scope of this conference.

ICSOFT 2008 received 296 paper submissions from more than 50 countries and all continents. To evaluate each submission, a double blind paper evaluation method was used: each paper was reviewed by at least two internationally known experts from the ICSOFT Program Committee. Only 49 papers were selected to be published and presented as full papers, i.e. completed work (8 pages in proceedings / 30' oral presentations). A total of, 70 additional papers describing work-in-progress were accepted as short papers for 20' oral presentation, leading to a total of 119 oral paper presentations. Another 40 papers were selected for poster presentation. The full-paper acceptance ratio was thus 16%, and the total oral paper acceptance ratio was 40%.

The papers you will find in this book were selected by the Conference Co-chairs and Program Co-chairs, among the papers actually presented at the conference, based on the scores given by the ICSOFT 2008 program committee members. We hope that you will find these papers interesting and that they provide a helpful source of reference in the future for all those who need to address any of the research areas mentioned above.

July 2009

José Cordeiro
Boris Shishkov
AlpeshKumar Ranchordas
Markus Helfert

Conference Committee

Conference Co-chairs

Joaquim Filipe   Polytechnic Institute of Setúbal / INSTICC, Portugal
AlpeshKumar Ranchordas   INSTICC, Portugal

Program Co-chairs

Markus Helfert   Dublin City University, Ireland
Boris Shishkov   University of Twente, / IICREST, The Netherlands

Organizing Committee

Paulo Brito   INSTICC, Portugal
Marina Carvalho   INSTICC, Portugal
Helder Coelhas   INSTICC, Portugal
Vera Coelho   INSTICC, Portugal
Andreia Costa   INSTICC, Portugal
Bruno Encarnação   INSTICC, Portugal
Bárbara Lima   INSTICC, Portugal
Vitor Pedrosa   INSTICC, Portugal
Vera Rosário   INSTICC, Portugal
Mónica Saramago   INSTICC, Portugal
José Varela   INSTICC, Portugal

Program Committee

Jemal Abawajy, Australia
Silvia Abrahão, Spain
Muhammad Abulaish, India
Hamideh Afsarmanesh, The Netherlands
Jacky Akoka, France
Markus Aleksy, Germany
Daniel Amyot, Canada
Tsanka Angelova, Bulgaria
Keijiro Araki, Japan
Alex Aravind, Canada
Farhad Arbab, The Netherlands
Cyrille Artho, Japan
Colin Atkinson, Germany
Rami Bahsoon, UK
Mortaza S. Bargh, The Netherlands
Joseph Barjis, USA

Bernhard Bauer, Germany
Bernhard Beckert, Germany
Noureddine Belkhatir, France
Fevzi Belli, Germany
Alexandre Bergel, France
Árpád Beszédes, Hungary
Maarten Boasson, The Netherlands
Wladimir Bodrow, Germany
Marcello Bonsangue, The Netherlands
Lydie du Bousquet, France
Mark van den Brand, The Netherlands
Manfred Broy, Germany
Gerardo Canfora, Italy
Cinzia Cappiello, Italy
Antonio Cerone, China
Sergio de Cesare, UK

Auxiliary Reviewers

Gabriela Aranda, Argentina
Tibor Bakota, Hungary
Vedran Batos, Croatia
Zoran Bohacek, Croatia
Philipp Bostan, Germany
Peter Braun, Germany
Patrick H.S. Brito, Brazil
Josip Brumec, Croatia
Stefano Busanelli, Italy
Glauco Carneiro, Brazil
Alejandra Cechich, Argentina
Yuan-Hao Chang, Taiwan
Che-Wei Chang, Taiwan
Shih-Chun Chou, Taiwan
Marcello Cinque, Italy
Methanias Colaço, Brazil
Daniela Soares Cruzes, USA
Katarina Curko, Croatia
Jörg Donandt, Germany
Ekaterina Ermilova, The Netherlands
Hua-Wei Fang, Taiwan
Y.Y. Fanjiang, Taiwan
Fausto Fasano, Italy
Ferenc Fischer, Hungary
Lajos Fülöp, Hungary
Ingolf Geist, Germany
Tamás Gergely, Hungary
Mehdi Golami, Denmark
Carmine Gravino, Italy
Nikola Hadjina, Croatia
Jie Hao, USA
Judith Hartmann, Germany
Wilke Havinga, The Netherlands
Christian Hofmann, The Netherlands
K.H. Hsu, Taiwan
Pi-Cheng Hsiu, Taiwan
Benjamin Hummel, Germany
Emilio Insfran, Spain
Judit Jász, Hungary

Tariq M. King, USA
Leonid Kof, Germany
Stephan Korsholm, Denmark
Pushpa Kumar, USA
Alessandro Lapadula, Italy
Stephane LoPresti, UK
Michele Loreti, Italy
Yung-Feng Lu, Taiwan
Andreas Lübcke, Germany
Paolo Medagliani, Italy
Kees van der Meer, The Netherlands
Maria de los Angeles Moraga, Spain
Simon S. Msanjila, The Netherlands
Joseph Okika, Denmark
Rocco Oliveto, Italy
Ignazio Passero, Italy
Viviane Malheiros de Pinho, Brazil
Daniel Ratiu, Germany
Sabine Rittmann, Germany
Giuseppe Scanniello, Italy
Eike Schallehn, Germany
Boris Shishkov, The Netherlands
István Siket, Hungary
Klaas Sikkel, The Netherlands
Bernd Spanfelner, Germany
Tom Staijen, The Netherlands
Pei-Lun Suei, Taiwan
Francesco Tiezzi, Italy
Gabriella Tóth, Hungary
Theocharis Tsigkritis, UK
László Vidács, Hungary
Boris Vrdoljak, Croatia
Vesna Bosilj Vuksic, Croatia
Yingbo Wang, USA
Feng Xie, China
N.L. Xue, Taiwan
Chuan-Yue Yang, Taiwan
Cheng Zhang, China
Chunying Zhao, USA

Invited Speakers

Colin Atkinson   Mannheim University, Germany
Dimitri Konstantas   University of Geneva, Switzerland
Michael Papazoglou   University of Tilburg, The Netherlands
Alexander Verbraeck   Delft University of Technology, The Netherlands

Table of Contents

Part IV: Information Systems and Data Management

Part V: Knowledge Engineering

User Defined Geo-referenced Information[*]

Dimitri Konstantas[1], Alfredo Villalba[1], Giovanna Di Marzo Serugendo[2],
and Katarzyna Wac[1]

[1] University of Geneva, Faculty of Social and Economic Sciences
24 rue General Dufour, CH-1204 Geneva, Switzerland
{dimitri.konstantas,alfredo.villalba,katarzyna.wac}@unige.ch
[2] Birkbeck, University of London, School of Computer Science and Information Systems
Malet Street, London WC1E 7HX, U.K.
dimarzo@dcs.bbk.ac.uk

Abstract. The evolution of technology allow us today to extent the "location based services" to fine grained services and allow any mobile user to create location based information and make it available to anyone interested. This evolution open the way for new services and applications for the mobile users. In this paper we present two novel mobile and wireless collaborative services and concepts, the Hovering Information, a mobile, geo-referenced content information management system, and the QoS Information service, providing user observed end-to-end infrastructure geo-related QoS information.

1 Introduction

From the dawn of time, people on the move were leaving marks (tags) in different places of interest serving different reasons, ranging from informing to other travellers regarding some important aspects of this specific location/point (ex. *"be careful, the rocks are unstable"* or *"water up the cliff"*), to simply filling their vanity (*"john was here"*), up to social/historical information (*"at this place the battle of Waterloo took place in 18 of June 1815"*).

Today with the rapid expansion and emergence of new wireless broadband networks, the wide adoption of mobile devices, like phones and PDAs and the recent integration of geo-localisation hardware (that is, GPS and the future Galileo receivers), a new form of geo-referenced tags is appearing placed not on physical objects, but in cyberspace. Spatial messaging, also called digital graffiti, air graffiti, or splash messaging, allows a user to publish a geo-referenced note so that other users reaching the same place will be able to received it and read it. Several applications and services are today available, allowing users to publish and access geo-referenced information, like it is done in the Google Earth Community where every user can post any geo-referenced information.

---

[*] The work presented in this paper is partially supported by the Swiss Secretariat for Education and Research (SER) under the COST project No. SER C08.0025 "Monitoring and analysis of wireless networks for a QoS-Prediction Service design and development", for the COST action IC0703, and the Swiss Federal Office for Professional Education and Technology (OFFT) under the project No CTI. 8643.2 PFES-ES "Geo-Tags".

J. Cordeiro et al. (Eds.): ICSOFT 2008, CCIS 47, pp. 1–8, 2009.

These types of *virtual* tags, combined with the shift in behaviour from mobile users who from information consumers have now become information and content creators, opens the way for novel collaborative mobile applications and services. The virtual tags paradigm can change the way we store, access and manage information, adding a new dimension, the location, to the information attributes.

In this paper we present two research ideas and projects we are working on, and which are based on the concept of virtual tags. The first is the concept of Hovering Information [1],[2] a mobile-content information management system. Hovering information is an attempt to understand how geo-localised information that is stored solely on mobile devices, can be managed. The second, is a down to earth service, the QoS Information Service, [3], [4], [5] aiming in providing detailed location based QoS information concerning the different available wireless networks, to mobile users. Both ideas are based on user collaboration and geo-referenced information.

2   Hovering Information

Today we can safely assume that future daily-life objects and people will be equipped with mobile devices, like RFIDs, environmental sensors and high quality mobile phones and PDAs, that will have large memory capacities and computing power, and will be inter-connected via high-capacity wireless networks.

Each person and object will thus be a node capable of generating, inserting, accessing, storing, processing or disseminating new types of active information. However, this information will no longer be stored in large fixed servers, as is the case today, but will be available only from the mobile nodes or daily-life objects, circulating and exchanged between the different users and objects. Considering the storage capacities available in mobile devices today (in the range of 1GB), and the rapidly increasing number of mobile wireless network supported devices (phones, PDAs, cameras, etc), the information stored in mobile devices will start becoming comparable to the information stored in static servers available on the wired internet, eventually creating a massively parallel and distributed information storage environment. In addition a new attribute will be attached to the information: its geo-location, that is the geographical location to where this is information is related or is relevant. Each user of mobile devices will carry with him a large amount of geo-referenced information becoming a potential mobile information server. However, based solely on *existing* approaches and models, this information will be inevitably confined in the mobile device itself; that is, the information will be associated to the mobile device that it is stored on, rather than its relevant location in space.

In order to allow this mobile information to make full use of the capabilities of the new mobile devices, a new information paradigm is needed, where the "information" is be able to detach itself from the physical media constrains (to which it is linked today) and associate itself with space and time. That is, the information should become an active and somewhat autonomous entity, independent of the ephemeral (mobile) hardware storage support; it should be able to *hover* from one mobile storage device to another in a quest to stay alive and within some specific location in space, or even define its migration plan moving independently of its hosts from one location to another. We call this type of active information, *Hovering Information*, and the location to which it is "assigned" the *anchoring location*.

The concept of Hovering Information provides a new paradigm for the dissemination, organization and access of information, integrating concepts from other research domains, ranging from sensor networks, to p2p, ad-hoc and autonomous networking, under a consistent model. The Hovering Information concept extends past networking concepts, from communication infrastructures to clustered massively parallel, self-regulated storing infrastructures, with the locus being the main information property, its relation (anchoring) and "obligation" to stay within a specific geographical location.

2.1 Overview of the Hovering Information Concept

The Hovering Information can be seen as an overlay on top of the P2P, ad-hoc and autonomous networks, linking information to a location and detaching the information from its physical support. Hovering Information can be shaped as text, image, video, alarm indication or any combination of these. A Hovering Information piece is linked to an anchoring point, which is determined by its geographical coordinates (longitude-latitude, or offset point from fixed positions or by any other method). The Hovering Information anchored to a point covers (i.e., can be logically be seen) a specific area around the anchoring point, that we call the anchoring area (or hot-spot). The anchoring area can be of any shape and can even represent a volume. Any user present in the anchoring area should be able to receive/access the information, irrespectively of its actual storage or originating host (fig. 1).

The creation of a hovering information service will provide mobile users a new concept and means for the creation and distribution of user-content, valorizing the ever-increasing power of their mobile devices. The fact the hovering information concept does not, per se, requires the existence of a mobile network operator, will open the way in the creation of new dynamic communication and social networking of users, based on geo-properties.

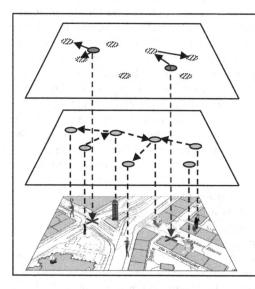

Hovering information provides an overlay network that is based on P2Pand Ad-hoc networks and is capable of linking information to a location rather than a device, by jumping from one device to another.

Add-hoc networks and P2P protocols allow the interconnection of user devices and the exchange of information that is anchored on the mobile device.

Users carrying mobile devices are moving in a city. Several fixed point devices, like phones, RFIDs etc provide points or reference and information.

Fig. 1. Overview of the hovering information concept

It will also allow the industry to create innovative health, entertainment and business applications such as health games to promote outdoor activities. Hovering information can be used as the "relay" that must remain alive, or moving from one hover cloud to another to score points (a cloud of geo referenced, self organizing, media and semantic rich content). This type of games can have an important impact on health related applications, like rehabilitation.

Other applications that can be developed on top of the hovering information include inter-vehicular geo-related information signaling without the need of fixed infrastructure, ephemeral location related information passing in high user-density areas (for example open air market, concerts etc - ad- hoc urban gatherings), hovering advertising and even in urban security allowing users to enter comments or warnings related to dangers in the urban environment, or even in disaster management.

Finally the use of Hovering Information can valorise the increasing but largely underused storage, processing capacity and resources of mobile devices. There are many millions of terabytes of storage capacity available in mobile devices that are hardly get any use. Hovering Information can provide the means to make a good case for this storage capacity and utilize the available resources of mobile phones.

2.2 Analysis of the Hovering Information Concept

Although the Hovering Information is a simple and elegant concept, it requires coordinated research at different levels and domains. We distinguish three main layers of research for the development of the concept and the models, and namely the **communication layer**, the **storage layer** and the **usage layer**.

At the **communication layer** the research issues relate to the recognition by the Hovering Information of the nearby hosts, the capabilities and potential for hovering from one host to another, the localization information and its accuracy etc. At this layer we build on research results coming from ad-hoc, p2p and autonomous networking, towards the understanding of the constraints and requirements for the communication between Hovering Information hosts. The main difference being that it is not the hosts that will define the policy of the communication, but the information itself (the paradigm is thus somehow reversed). For this we need to define the required communication primitives that will be available to the higher layers.

At the **storage layer** we concentrate on issues related to the survivability of the Hovering Information pieces. Based on the primitives of the communication layer, a Hovering Information piece controls when, how and where will it hover from one host to another. For this part of research we start be evaluating models coming from different areas (like bio-models) and understand how a new model can be designed to suit the Hovering Information needs. At the storage level each Hovering Information piece is seen independently of any others.

At the **usage layer** the Hovering Information pieces are seen as a whole. In this layer a Hovering Information piece will be a part of a "community" of Hovering Information pieces serving the mobile users. In the usage layer issues related to the capabilities of the nodes, the multiplicity and complementarities of Hovering Information pieces will be studied. Questions related to the meta-information carried by the Hovering Information pieces, the need or not of classification of Hovering Information, the methods to identify available information at different hosts in the anchoring location with minimal overhead as well as considerations of the local host preferences (like power and capacity considerations), will be some of the research issues in this layer.

The development of the Hovering Information concept requires the clarification and study of many issues and problems, which are new in the area of wireless information management. In order to demonstrate the complexity of the issues related to the abstract concept, let us take a simple example. We assume that we have a small number of roaming mobile users with a simple sensor measuring the temperature of the air and making it available to other users, as Hovering Information pieces, linked to the location where the temperature was sampled. Other users will be able to be informed of the temperature at a location, even if they do not posses the required sensor. A temperature at a location will change over the time and different temperature-capable mobile users will pass-by a specific location during the day. The questions that arise are many; when more than one measurements of the temperature are available at the location, which is one is valid? Time stamps can solve this problem, but then what happens to the old measurements? Can a new measurement delete an old one? And what about the need to keep track of the temperature changes during the day? How a mobile user entering the anchoring area can "read" the latest defined temperature, without having to receive all measurements from time immemorial and choose the latest? How oldest readings will decide that they are no longer valid and die-away? How can they know that there is a new reading available and that no-one is interested in the historical data or that other copies exist in the area? How can a user receiving the temperature as Hovering Information can trust the reading as coming from a reliable source?

3   End-to-End QoS-Predictions' Information Service

Mobile wireless communications are becoming the basis for future applications and services, which will be Context Aware (CA), able to adapt according to the changing user needs and location. Our vision is that the forthcoming generation of mobile CA services and applications will be based on systems able to collect data from a multitude of sources in the user's environment and processes them for creating a model of the user's world and identify patterns and context in the user's daily activities. Based on these patterns and context, the user services and applications will adapt themselves in order to provide the best possible service quality. One of these context elements that is relevant to mobile users is the quality of the wireless networks, which is expressed as the Quality of Service (QoS) of the network.

The question arises on how a mobile service with particular user quality requirements can be aware of the QoS provided by diverse underlying wireless networks at a given user's context, and this in order to facilitate the service and application's choice of (best) network to connect. The today available data regarding the QoS provided by public and private wireless network operators are not sufficient for the future highly demanding applications, since they can only give a very coarse level of QoS, based only on antenna related information (power) and unable to take into consideration obstacles, numbers of users served, show areas etc. The required fine grain QoS required is out of reach with existing solutions.

The QoS prediction Information Service we propose is based on Traffic Monitoring and Analysis (TMA) methodologies for wireless networks. Our idea is to collect QoS data as context, measured by the users of mobile services at different times and locations. Based on the collected (historical) data and using TMA methodologies we will be able to create a QoS-predictions' service able to provide fine grained end-to-end QoS information regarding wireless networks available at a location and time.

3.1 Basic Assumptions

Our approach is based on a set of realistic (in our view) assumptions. The first assumption is that at any location and time various connectivity possibilities (can) coexist and are accessible, offering different QoS to the user. Although this is possible in some cases today, current mobile applications are unaware and unable to make use of these offers due to the user 'lock-in' restrictions imposed by MNOs.

Our second assumption is that the user's QoS requirements change during the lifetime of the service delivery, depending on the user's context change. This assumption is valid especially for services that have a long duration, like for example continuous mobile health monitoring service, where the required-QoS will change along the health state of the user. In case of emergency, i.e. epileptic seizure, the required-QoS will be stricter (i.e. shorter delays required), than in case of a non-emergency situation.

Our third assumption is that a mobile user is able to handover between 'any' one of the available networks, in order to obtain the best-of the end-to-end offered-QoS fulfilling his required-QoS.

Our fourth assumption is that it is not possible to negotiate the end-to-end offered-QoS over the heterogeneous communication path; as a consequence it is possible to obtain only the best-effort end-to-end offered-QoS.

Based on the above assumptions we make two major claims, which define the context of our research. Our first claim is that the mobile service infrastructure must not only be context-aware, but also QoS-aware; and that the availability of information about these offered-QoS will result in user's intelligent choice of connectivity and an intelligent adaptation of application protocol parameters.

3.2 Overview of the Concept and Service

In order to demonstrate the main idea we provide an example of usage scenario. In our scenario, the communication infrastructure supports data exchange between a mobile patient (carrying a mobile vital signals' monitoring device) and a healthcare provider in the healthcare centre. The communication infrastructure is hybrid and it is delivered by different entities. In general, the communication infrastructure consists of at least two (interacting) networks belonging to at least two organizations. What is important for any application and service, is not just the QoS of the wireless part but the complete end-to-end QoS of the communication path.

In figure 2 a mobile service user moves along the trajectory from the point A to B. The mobile service is delivered to the user from a backend system placed in point C. There are wireless communication networks of multiple operators available along the user's trajectory and the QoS aware service platform has already collected many QoS data from other users.

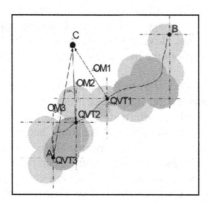

Fig. 2. User trajectory and network access

At point A, the user service requests a QoS information regarding the available networks. It occurs that there are two networks available in this location, from two different operators. Based on the provided available QoS information, the user chooses one operator (red), which provides a service with a quality matching to the user requirements. The user also posts his observer QoS performance measurements (QM) in the form of virtual tags containing observed QoS -information (denoted as QoS Virtual Tags (QVT)). This is repeated throughout the complete trajectory of the user from A to B. The information posted by the users is used by the QoS-IS to calculate estimates regarding the expected QoS at a specific location and time, eventually enabling the intelligent choice of the network technology and operator to use at a particular location.

In order to be able to correctly use the collected QMs, different factors need to be taken into account. The QoS-IS, will collect a very large number of QMs coming from different users, at different locations, times, applications networks etc. However, we must judge the validity of the QMs, their importance and their accuracy, before using them for QoS predictions. In other words, we must define how much we trust a specific QM. There may be many reasons for distrusting a QM: one can be the fact that the user device creates a bottleneck due to low processing power and thus the measured performance is not correct; another. can be that that the QM is "too old" and the actual state of the network has changed ever since, or even that the specific QM was taken under bad environmental conditions (like strong rain) and cannot be applied when these conditions are not longer present. Finally an important question will be how extrapolate the QoS for a specific location, based on measurements of nearby locations. Nevertheless we argue that partial information might worth more for a user than no information at all.

4 Conclusions

The wide availability and the seamless integration of geo-localisation hardware into every-day devices, will open the way for the development of new context and location aware services and applications. In addition user collaborative mobile platforms and the need for the casual exchange of geo-referenced information will revolutionise the way mobile information is managed and accessed. In this paper we presented two running research projects, the Hovering Information and the QoS Information Service, that are based on collaborative mobile user information exchange and high precision geo-localisation.

The two projects introduce a shift in the paradigm for managing and using information. The Hovering Information introduces a new layer at the information management level for location related information. Information is no longer bound to the physical storage but rather to the physical location, becoming independent of the "ephemeral" and temporary mobile storage means. This shift of paradigm will require further research for resolving questions like security, persistency of information etc.

The QoS-Information Service on the other hand is a down to earth service that empowers the mobile users in their selection of wireless network providers. Whereas today the users are left on their own to choose the wireless network, the QoS-IS will allow them to make an intelligent selection based on concrete criteria, like the observed wireless network QoS. Clearly in order to have a successful QoS information service, the users should be able to seamlessly and without interruption "roam" from one network to another and from one provider to another, something which today is not possible, for both technical and (more important) commercial reasons (customer locking to a wireless network). We expect that within the next decade the commercial restrictions will be eventually solved (by law or for competition reasons). Our vision is that within the early 2010's mobile devices will be able to automatically, transparently and seamlessly interconnect and exchange geo-referenced (among others) information.

References

1. Castro, A.V., Di Marzo Serugendo, G., Konstantas, D.: Hovering Information - Self-Organising Information that Finds its Own Storage. In: Proceeding of the IEEE International Conference on Sensor Networks, Ubiquitous, and Trustworthy Computing (SUTC 2008), Taichung, Taiwan (June 2008)
2. Di Marzo Serugendo, G., Villalba, A., Konstantas, D.: Dependability Requirements for Hovering Information. In: Workshop on Dependable Application Support for Self-Organizing Networks, DSN 2007, The 37th Annual IEEE/IFIP International Conference on Dependable Systems and Networks, Edinburgh International Conference Centre, Edinburgh, UK, June 25-28 (2007)
3. Wac, K., van Halteren, A., Bults, R., Broens, T.: Context-aware QoS pro-visioning for an M-health service platform. International Journal of Internet Protocol Technology 2(2), 102–108 (2007)
4. Broens, T., van Halteren, A., van Sinderen, M., Wac, K.: Towards an application framework for context-aware m-health applications. International Journal of Internet Protocol Technology 2(2), 109–115 (2007)
5. Wac, K., Arlos, P., Fiedler, M., Chevul, S., Isaksson, L., Bults, R.: Accuracy evaluation of application-level performance QoS measurements. In: Next Generation Internet Net-works, Trondheim, Norway, May 21-23, pp. 1–5. IEEE Computer Society Press, Los Alamitos (2007)

Extending the SSCLI to Support Dynamic Inheritance

Jose Manuel Redondo[1], Francisco Ortin[1], and J. Baltasar Garcia Perez-Schofield[2]

[1] University of Oviedo, Computer Science Department, Calvo Sotelo s/n, 33007, Oviedo, Spain
`redondojose@uniovi.es, ortin@lsi.uniovi.es`
[2] University of Vigo, Computer Science Department, As Lagoas s/n, 32004, Ourense, Spain
`jgarcia@uvigo.es`

Abstract. This paper presents a step forward on a research trend focused on increasing runtime adaptability of commercial JIT-based virtual machines, describing how to include dynamic inheritance into this kind of platforms. A considerable amount of research aimed at improving runtime performance of virtual machines has converted them into the ideal support for developing different types of software products. Current virtual machines do not only provide benefits such as application interoperability, distribution and code portability, but they also offer a competitive runtime performance.

Since JIT compilation has played a very important role in improving runtime performance of virtual machines, we first extended a production JIT-based virtual machine to support efficient language-neutral structural reflective primitives of dynamically typed programming languages. This article presents the next step in our research work: supporting language-neutral dynamic inheritance for both statically and dynamically typed programming languages. Executing both kinds of programming languages over the same platform provides a direct interoperation between them.

1 Introduction

Dynamically typed languages like Python [1] or Ruby [2] are frequently used nowadays to develop different kinds of applications, such as adaptable and adaptive software, Web development (the *Ruby on Rails* framework [3]), application frameworks (*JSR 223* [4]), persistence [5] or dynamic aspect-oriented programming [6]. These languages build on the Smalltalk idea of supporting reasoning about (and customizing) program structure, behavior and environment at runtime. This is commonly referred to as *the revival of dynamic languages* [7].

The main objective of dynamically typed languages is to model the dynamicity that is sometimes required in building high context-dependent software, due to the mobility of both the software itself and its users. For that reason, features like meta-programming, reflection, mobility or dynamic reconfiguration and distribution are supported by these languages. However, supporting these features can cause two main drawbacks: the lack of static type checking and a runtime performance penalty.

Our past work [8] [9] [10] [11] has been focused on applying the same approach that made virtual machines a valid alternative to develop commercial software. Many virtual machines of dynamically typed languages are developed as interpreters. Designing a JIT compiler for a dynamically typed language is a difficult task. So, our approach

J. Cordeiro et al. (Eds.): ICSOFT 2008, CCIS 47, pp. 9–22, 2009.

uses an existing commercial virtual machine JIT compiler to evaluate whether it is a suitable way to achieve advantages like improving their runtime performance. The chosen virtual machine was the Microsoft Shared Source implementation of the .Net platform (SSCLI aka Rotor).

Nowadays, there are initiatives to support dynamically typed languages modifying a production JIT-based virtual machine, such as [12] or [13]. Our approach also adds new features to an existing virtual machine, but we want to introduce full low-level support for the whole set of reflective primitives that support the dynamicity of these languages, as part of the machine services. The main advantages of our approach are:

1. Language processors of dynamically typed languages can be implemented using the modified machine services directly. Low-level support of structural reflection eases the implementation of its dynamic features.
2. Full backwards compatibility with legacy code. Instead of modifying the syntax of the virtual machine intermediate language, we extended the semantics of several instructions.
3. Reflective primitives are offered to any present or future language (they are language - neutral). As a consequence, no existing language will need to be syntactically changed to use them.
4. Interoperability is now possible between static and dynamically typed languages, since the machine supports both of them.
5. Obtain a performance advantage, since it is not necessary to generate extra code to simulate dynamic functionalities –see Sect. 2.2.

Existing implementations do not offer all the advantages and features of our approach. For example, the Da Vinci virtual machine [14] (prototype implementation of the JSR292 specification [12]) will likely add new instructions to its intermediate language. By doing this, its new features will not be backward compatible, since compilers must be aware of these instructions to benefit from the extended support they provide. Another example is the Microsoft DLR [13], which aims to implement on top of the CLR several features to support dynamically typed languages. Languages must be designed to target the DLR specifically to benefit from its features, so this implementation is not language-neutral.

In our past work we had successfully implemented most of these primitives into the machine, enabling low-level support to add, modify or delete any object or class member. Prototype-based object-oriented semantics [15] were also appropriately introduced to solve those cases in which the existing class-based model had not enough flexibility to allow us to easily implement the desired functionality. The original machine object model evolved into a hybrid model that support both kinds of languages without breaking backwards compatibility, also allowing direct interoperability as we mention in Sect. 1 [9]. The research prototype that incorporates all these dynamic features is named Reflective Rotor or ЯROTOR.

In this paper we will explain the next step of our work, designing and implementing a new reflective primitive into our system which will enable the users to effectively use single dynamic inheritance over the hybrid object model that ЯROTOR now supports. Languages that support this feature (such as Python [1]) make possible, for example, to change the inheritance hierarchy of a class at run time [16], allowing a greater degree of flexibility to its programs. With the addition of this new primitive, the full set of structural reflective primitives will be low-level supported by ЯROTOR.

The rest of this paper is structured as follows. Section 2 describes the background of our work, explaining several concepts involved in its development. Section 3 details the design of the dynamic inheritance support over the existing machine. Section 4 presents details about the implementation of our design and finally Sect. 5 and 6 describe the current state of the project, the final conclusions, and future lines of work.

2 Background

2.1 Type Systems of Dynamically Typed Languages

Dynamically typed languages use dynamic type systems to enable runtime adaptability of its programs. However, a static type system offers the programmer the advantage of early detection of type errors, making possible to fix them immediately rather than discovering them at runtime or even after the program has been deployed [17]. Another drawback of dynamic type checking is low runtime performance, discussed in the Sect. 2.2. There are some research works that try to partially amend the disadvantages of not being able to perform static type checking. These include approaches like integrating unit testing facilities and suites with dynamically typed languages [18] or allowing static and dynamic typing in the same language [19] [20]. This is the reason why we have focused our efforts in improving runtime performance.

2.2 Runtime Performance of Dynamically Typed Languages

Looking for code mobility, portability, and distribution facilities, these languages are usually compiled to the intermediate language of an abstract machine. Since its computational model offers dynamic modification of its structure and code generation at runtime, existing virtual machines of dynamically typed languages are commonly implemented as interpreters. An interpreter can easily support its dynamic features with a lower development cost if compared with an equivalent JIT compiler. This fact, plus the runtime type checking additional cost, can cause a performance penalty on average, if compared to "static" languages.

Since the research in *customized dynamic compilation* applied to the Self programming language [21], virtual machine implementations have become faster by optimizing the binary code generated at run time using different techniques. An example is the dynamic adaptive *HotSpot* optimizer compilers. Speeding up the application execution of dynamically typed languages by using JIT compilation facilitates their inclusion in commercial development environments.

Most recent works that use commercial virtual machine JIT compilers to improve runtime performance of dynamically typed languages are aimed to the Java and .NET platforms. Several compilers have been developed that generate Java or .Net bytecodes. These machines do not support structural reflection, as they were created to support static languages. Generating extra code is then needed to simulate dynamic features over these machines, leading to a poor runtime performance [22]. Examples of this approach are *Python for .Net* from the Zope Community, *IronPython* from Microsoft or *Jython* for the JVM platform.

As opposed to the previous approach, we use a virtual machine with JIT compilation to directly support any dynamically typed language. Unlike existing implementations, we extended the static computational model of an efficient virtual machine, adding the reflective services of dynamically typed languages. This new computational model is then dynamically translated into the native code of a specific platform using the JIT compiler. So, instead of generating extra code to simulate the computational model of dynamically typed languages, the virtual machine supports these services directly. As a result, a significant performance improvement is achieved -see Sect. 2.4.

2.3   Structural Reflection

Reflection is *the capability of a computational system to reason about and act upon itself, adjusting itself to changing conditions* [23]. In a reflective language, the computational domain is enhanced with its own representation, offering its structure and semantics as computable data at runtime. Reflection has been recognized as a suitable tool to aid the dynamic evolution of running systems, being the primary technique to obtain the meta-programming, adaptiveness, and dynamic reconfiguration features of dynamically typed languages [24].

The main criterion to categorize runtime reflective systems is taking into consideration what can be reflected. According to that, three levels of reflection can be identified: *Introspection* (read the program structure), *Structural Reflection* (modify the system structure, adding fields or methods to objects or classes, reflecting the changes at runtime) and *Computational (Behavioral) Reflection* (System semantics can be modified, changing runtime behavior of programs).

2.4   ЯRotor

As was previously mentioned, compiling languages to the intermediate code of a virtual machine offers many benefits, such as platform neutrality or application interoperability [25]. In addition, compiling languages to a lower abstraction level virtual machine improves runtime performance in comparison with direct interpretation of programs.

Therefore, we have used the Microsoft .Net platform as the targeted virtual machine to benefit from its design, focused on supporting a wide number of languages [26] [27]. The selection of an specific distribution was based on the necessity of an open source implementation to extend its semantics and an efficient JIT compiler. The SSCLI (Shared Source Common Language Infrastructure, aka Rotor) implementation has been our final choice because it is nearer to the commercial virtual machine implementation: the Common Language Runtime (CLR) [28].

To date, we have successfully extended the execution environment to adapt the semantics of the abstract machine, obtaining a new reflective computational model that is backward compatible with existing programs. We have also extended the Base Class Library (BCL) to add new structural reflection primitives (add, remove and modify members of any object or class in the system), instead of defining new IL statements. We refer to this new platform as Reflective Rotor or ЯRotor.

The assessment of our current ЯROTOR implementation has shown us that our approach is the fastest if compared to widely used dynamically typed languages such as Python and Ruby [9]. When running reflective tests, our system is on average almost 3 times faster than the fastest commercial dynamically typed languages implementations. When running static code, our system is 3 times faster than the rest of tested implementations, needing a significantly lower memory usage increment (at most 102% more). Finally, we have also evaluated the cost of our enhancements. When running real applications that do not use reflection at all, empirical results show that the performance cost is generally below 50%, using 4% more memory.

2.5   The Object Model of ЯROTOR

The current object model of our machine is the result of the needed evolution of the original SSCLI class-based model to give proper and efficient support to structural reflection primitives. In the class-based model, every instance must have a corresponding class defining all its attributes and methods (state and behavior). Structurally reflective operations must not interfere with this rule. Class-based languages that allow structural changes within its classes [29][30] update the structure of every instance of a modified class and its subclasses to reflect these changes. This mechanism was defined as a type re-classification operation in class-based languages, changing the class membership of an object while preserving its identity [31][32]. This mechanism has also been referred to as schema evolution in the database world [33].

The class-based model rules make single instance evolution much more difficult. Since any object must follow the structure of its corresponding class, changing the structure of a single instance without breaking the model rules is not straightforward. This problem was detected in the development of MetaXa, a reflective Java platform implementation [33]. The approach they chose was the same as the one adopted by some object-oriented database management systems: schema versioning. A new version of the class (called *shadow* class in MetaXa) is created when one of its instances is reflectively modified. This new class is the type of the recently customized object.

Shadow classes in MetaXa caused different problems such as maintaining the class data consistency, class identity or using class objects in the code, involving a really complex and difficult to manage implementation [33]. A conclusion of the MetaXa research project was that the class-based OO model does not fit well in structural reflective environments. This is why we introduced the prototype-based model into ЯROTOR.

In the prototype-based object-oriented computational model the main abstraction is the object, suppressing the existence of classes [15]. Although this computational model is simpler than the one based on classes, any class-based program can be translated into the prototype-based model [34]. In fact, this model has been considered as a universal substrate for object-oriented languages [35].

For our project, the most important feature of the prototype-based OO computational model is that it models structural reflection primitives in a consistent, easy and coherent way [9]. Dynamically typed languages use this model for the same reason. Modifying the structure of a single object is performed directly, because any object maintains its own structure and even its specialized behavior. Shared behavior could be placed in the so called trait objects, so its customization implies the adaptation of types.

Although the so called *Common Language Infrastructure* (CLI) tries to support a wide set of languages, the .Net platform only offers a class-based object-oriented model optimized to execute "static" languages. In order to allow prototype-based dynamically typed languages to be interoperable with any existing .Net language or application, and to maintain backward compatibility, we supported both models. This way we can run static class-based .Net applications and dynamic reflective programs. .Net compilers could then select services of the appropriate model depending on the language being compiled. Consequently, compilers for a wider range of languages could be implemented. These compilers will generate the intermediate code of our machine, that is in fact syntactically the same as the original machine (to maintain backward compatibility). Examples of these languages are:

1. Class-based languages with static typing (C#).
2. Prototype-based languages with dynamic typing (e.g. Python [1]).
3. Class-based languages with static and dynamic typing (e.g. Boo [36] or StaDyn [20]).
4. Prototype-based languages with static typing (e.g. StrongTalk [37]).

2.6   Dynamic Inheritance

When passing messages to a particular object, conventional class-based languages use a concatenation-based inheritance strategy. This means that all members (either derived or owned) of a particular class must be included in the internal structure of this class. Using this approach enables compile-time type checking. This will prove that there can never be an error derived from invoking a non existing message [38], but at the expense of flexibility. In contrast, dynamically typed languages use delegation-based inheritance mechanism, iterating over the hierarchy of an object searching for the intended member. Since our system introduced prototype-based semantics, we also implemented a delegation-based inheritance mechanism to be used together.

Delegation - based inheritance is complemented with dynamic inheritance. In contrast with conventional class-based languages, prototype-based languages allow a inheritance hierarchy to be changed at run time [16]. More specifically, dynamic inheritance refers to the ability to add, change or delete base classes from any class at run time. It also includes the ability to dynamically change the type of any instance. This results in a much more flexible approach, allowing objects and classes of any program to better adapt to changing requirements. This type of inheritance is implemented by languages such as Python [1], Self [21], Kevo [39], Slate [40] or AmbientTalk [41].

Therefore, it is necessary to create the means to provide adequate support for dynamic inheritance to give a complete support of the dynamic features of dynamically typed languages.

3   Design

In this section we will analyze the semantics of dynamic inheritance when applied over the object model of яRotor. Therefore, we must take into account both class-based and prototype-based semantics. It is important to state that, while either models are supported, they will not be both present at the same time. Languages could be either class-based or prototype-based, but they will not use both models together. The implementation of dynamic inheritance will be done creating a new *setSuper* primitive, which will be added to the already existing ones. We consider two main operations over each object model:

1. **Instance type change** (*setSuper(instance, type)*): Substituting the current type of an instance with another one, performing the appropriate changes on its structure to match the new type.
2. **Class inheritance tree change** (*setSuper(type, type)*): Substituting the base type of a class with another one, performing the necessary changes over the class hierarchy and instances.

In order to give a precise description, we will formalize the behavior of the new *setSuper* primitive. We assume that:

1. C_a is the attribute set of class C.
2. C_m is the method set of class C.
3. C_p is the member set of class C ($C_p = C_a \cup C_m$)
4. Once these elements are defined, the full set of attributes accessible from C can be calculated as:
$$C_a^+ = C_a \cup D_a^+, \forall D \text{ superclass of } C \tag{1}$$
5. At the same time, the full set of methods accessible from C is:
$$C_m^+ = C_m \cup D_m^+, \forall D \text{ superclass of } C \tag{2}$$
6. So, all the members of a particular class C are calculated this way:
$$C_p^+ = C_a^+ \cup C_m^+ \tag{3}$$

The described formalizations have been designed without breaking the restrictions of the object model over which they are applied. This design also relies on the reflective primitives that were implemented in our previous work –see Sect. 2.4. Each formalization is given an example to clarify its functionality. All the given examples will refer to the class diagram shown in Fig. 1.

3.1 Class-Based Model

The design of an **instance type change** is formalized as follows. Fig. 2 shows an example of this operation:

1. Let X, Y be classes and $o: X$.
2. Using (3), the *setSuper(o, Y)* primitive call modifies o structure this way:
   (a) Delete from o the member set D:
$$D = X_p^+ - (X_p^+ \cap Y_p^+) \tag{4}$$

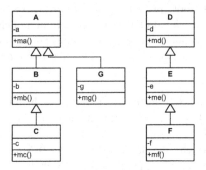

Fig. 1. Example class diagram

Fig. 2. Type change (class-based model)

(b) Add to *o* the member set *A*:

$$A = Y_p^+ - X_p^+ \tag{5}$$

The design of an **inheritance tree change** of a class is formalized as follows. Fig. 3 shows an example of this operation:

1. Let *X, Y* be classes. Let *Z* be the base class of *X*.
2. Using (3), the *setSuper(X, Y)* primitive call modifies class *X* structure this way:

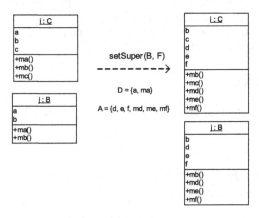

Fig. 3. Inheritance tree change (class-based model)

(a) Delete from X the member set D:

$$D = Z_p^+ - (Z_p^+ \cap Y_p^+) \qquad (6)$$

(b) Add to X the member set A:

$$A = Y_p^+ - Z_p^+ \qquad (7)$$

It should also be noted that instance members of modified classes are dynamically updated when they are about to be used (lazy mechanism). The existing primitives already use this mechanism [9], so we will take advantage of it. This is much faster than actively modifying all instances when a change is performed, specially if a large number of them are present.

3.2 Prototype-Based Model

The design of an **instance type change** is formalized as follows. Fig. 4 shows an example of this operation:

1. Let X, Y be classes and o: X.
2. Using (2), the *setSuper(o, Y)* primitive call modifies o structure this way:
   (a) Delete from o the method set D:

$$D = X_m^+ - (X_m^+ \cap Y_m^+) \qquad (8)$$

   (b) Add to o the method set A:

$$A = Y_m^+ - X_m^+ \qquad (9)$$

It should also be noted that existing attributes are always maintained in the instances (no attribute is added nor deleted).

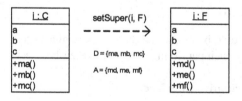

Fig. 4. Type change (prototype-based model)

The design of an **inheritance tree change** of a class is formalized as follows. Fig. 5 shows an example of this operation:

1. Let X, Y be classes. Let Z be the base class of X.
2. Using (2), the *setSuper(X, Y)* primitive call modifies class X structure this way:
   (a) Delete from X the method set D:

$$D = Z_m^+ - (Z_m^+ \cap Y_m^+) \qquad (10)$$

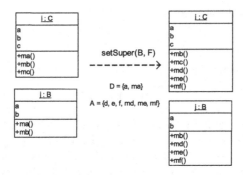

Fig. 5. Inheritance tree change (prototype-based model)

(b) Add to X the method set A:

$$A = Y_m^+ - Z_m^+ \tag{11}$$

The only difference between both models formalization is that the prototype-based model maintain instance attributes. Finally, it must be remarked that all operations will comply the following assumptions:

1. The computational complexity of any operation is $O(n)$, being n the number of classes present in the X, Y hierarchies. The algorithm that implement (1)(2) goes through each class once to perform the requested operation.
2. In ЯROTOR, when we ask via introspection for the type of an object or the base type of a class, it must respond with the dynamically assigned type.
3. Adding a member to a set that already contains it has no computation.
4. When new attributes are incorporated to instances or classes as a consequence of any of the described operations, default values are assigned according to the value that was assigned to them in its original declaration.

4   Implementation

The implementation could be divided in three important aspects. We will describe the techniques followed in each part to implement the described design.

4.1   The *setSuper* Primitive Interface

As we made with the primitives to add, modify and remove members, we added to our *NativeStructural* class the new *setSuper* primitive. The virtual machine core is then extended to incorporate this new service, linking the primitive call interface to its low-level implementation [9]. Therefore, our new extension naturally integrates with the already created infrastructure, being part of the virtual machine services. The simplest way to distinguish what object model is going to be used is adding a third boolean parameter to the primitive. This way, a user can easily choose either model and the system will select the appropriate behavior. All the virtual machine changes we will present were possible because the original virtual machine static type system has been "relaxed" as a consequence of our past work.

4.2   Instance Manipulation

One of the most challenging tasks in this work is how to effectively translate the designed dynamic inheritance model into the machine internals. The new *setSuper* primitive will take advantage of the primitives added in our previous work to successfully implement its features. However, these primitives never needed to modify parent-child or instance-class relationships, so this concrete part of the system must be studied. The capability to add or remove members to any object or class in the system fell short to implement the desired functionality. Although the changed entity could have the exact interface we pretend, its internal type will not be correct.

So, in order to achieve an instance type change, we have to carefully review the low-level infrastructure representation of instances [28]. This way, we found that every instance is bound to its class using a pointer to a unique class: the *MethodTable*. We found that dynamically changing an instance method table produces the desired effect, and the system responds as expected when types of instances are dynamically requested.

Therefore, by appropriately combining our existing structural reflection primitives with a *MethodTable* substitution of the involved instance we can achieve the desired type change functionality. Once this is solved, it was easy to compute the A and D member sets and apply them to implement the described model design, using the primitives of our previous work [9].

4.3   Class Manipulation

Although in the internal structure of a class there exists a pointer to its base type, dynamically changing this pointer to the desired class would not work. This is because the original *SSCLI* virtual machine implements introspection routines that interacts directly with internal structures of classes to obtain members and other requested data. It was found that this introspection code uses several built-in integrity tests and internal checking that produced wrong behavior if we change the aforementioned pointer. These checks are also invoked in several parts of the system to maintain class integrity, so it turned impractical to modify them. So, in order to create the desired functionality, a workaround solution must be used.

The implementation of the reflective primitives in our previous work forced us to use an auxiliary *SSCLI* class in order to store the information we need [9]. This element is called the *SyncBlock*. Every object and class in the system have one attached to its internal structure. We use this *SyncBlock* to our advantage in order to implement the desired functionality and solve the aforementioned implementation problems.

By storing a pointer to the dynamically assigned base class in this *Syncblock*, we can modify the implementation of the desired introspection primitives without modifying the "real" base type pointer. This way, any introspection primitive implementation in the system (*Type.getMethods()*, *Type.getFields()*, *Object.GetType()*, ...) could be modified to produce correct results, taking into account the dynamically assigned base class only if it is present. By using this approach we can use the original virtual machine code if no change is performed (more performance) and the internal checks will not report errors.

5   Current State of the Implementation

Our current implementation already supports dynamic inheritance. The two operations mentioned in Sect. 3 have been incorporated in the new *setSuper* method added to the *NativeStructural* class of the *System.Reflection.Structural* namespace. This support is offered in a language-neutral way, so any language that targets the virtual machine could use it. We present two examples showing the usage and behavior of the implemented functionality. They follow the design, formalizations and samples presented in Sect. 3. This is an example of changing the type of an instance:

```
C i = new C(); F f = new F();
//Type change of instance i, class model.
NativeStructural.setSuper(i, typeof(F), true);
//  - i Methods: md, me, mf, GetType, ToString, Equals, GetHashCode
//  - Attributes of i: d, e, f
//Return to the C class (another type change in the class model).
NativeStructural.setSuper(i, typeof(C), true);
//  - i Methods: ma, mb, mc, GetType, ToString, Equals, GetHashCode
//  - Attributes of i: a, b, c
```

This is an example of changing the inheritance tree of a class. Note that all instances of the class and its subclasses are affected by these changes.

```
C i = new C(); B j = new B();
//Inheritance tree change, prototype model
NativeStructural.setSuper(typeof(B), typeof(F), false);
//  - i Methods: mb, mc, md, me, mf, GetType, ToString, ...
//  - Attributes of i: a, b, c
//  - j Methods: mb, md, me, mf, GetType, ToString, ...
//  - Attributes of j: a, b
```

6   Conclusions

The major contribution of our work is the design and implementation of language-neutral dynamic inheritance support over a production high-performance JIT-based virtual machine. Nowadays, there is an active research line whose aim is to give more support to dynamically typed languages using JIT-based virtual machines [14] [13]. Our system can be classified into this research line, trying a different approach to attain these objectives. Our dynamic inheritance support is combined with the already implemented reflective primitives, offering complete support to structural reflection fully integrated into the machine internals. This approach offers the following benefits:

1. Any language (class or prototype-based) could benefit from structural reflection without altering its specifications.
2. Low-level support of all the structural reflection primitives into the virtual machine, allowing any dynamically typed language to be completely implemented over its services, without adding any extra abstraction layer to simulate dynamic features.

3. Extend the interoperability present in the original machine to include dynamically typed languages, enabling them to interoperate with static ones.
4. Full backward compatibility with legacy code, since the intermediate code of the machine is not syntactically changed.
5. Due to the commercial character of the SSCLI, it is possible to directly offer its new services to existing frameworks and languages designed to work with it.

Future work will incorporate meta-classes to the machine, taking advantage of the existing dynamic features. We are also developing a language that supports both dynamic and static typing, making the most of our ЯROTOR implementation [20].

Acknowledgements. This work has been partially funded by *Microsoft Research* with the project entitled *Extending Rotor with Structural Reflection to support Reflective Languages*. It has also been partially funded by the *Department of Science and Innovation* (Spain) under the *National Program for Research, Development and Innovation*, project *TIN2008-00276*. Our work is supported by the *Computational Reflection research group* (http:// www.reflection.uniovi.es) of the University of Oviedo (Spain). The materials presented in this paper are available at http://www.reflection.uniovi.es/rrotor.

References

1. Rossum, G.V., Drake, F.L.: The Python Language Ref. Manual. Network Theory (2003)
2. Thomas, D., Fowler, C., Hunt, A.: Programming Ruby, 2nd edn. Addison-Wesley, Reading (2004)
3. Thomas, D., Hansson, D.H., Schwarz, A., Fuchs, T., Breed, L., Clark, M.: Agile Web Development with Rails. A Pragmatic Guide. Pragmatic Bookshelf (2005)
4. Grogan, M.: JSR 223. scripting for the Java platform (2008), http://www.jcp.org
5. Ortin, F., Lopez, B., Perez-Schofield, J.B.: Separating adaptable persistence attributes through computational reflection. IEEE Soft. 21(6) (November 2004)
6. Ortin, F., Cueva, J.M.: Dynamic adaptation of application aspects. Journal of Systems and Software (May 2004)
7. Nierstrasz, O., Bergel, A., Denker, M., Ducasse, S., Gaelli, M., Wuyts, R.: On the revival of dynamic languages. In: Software Composition 2005. LNCS. Springer, Heidelberg (2005)
8. Ortin, F., Redondo, J.M., Vinuesa, L., Cueva, J.M.: Adding structural reflection to the SSCLI. Journal of Net Technologies, 151–162 (May 2005)
9. Redondo, J.M., Ortin, F., Cueva, J.M.: Optimizing reflective primitives of dynamic languages. Int. Journal of Soft. Eng. and Knowledge Eng (2008)
10. Redondo, J.M., Ortin, F., Cueva, J.M.: Diseño de primitivas de reflexión estructural eficientes integradas en SSCLI. In: Proceedings of the JISBD 2006 (October 2006)
11. Redondo, J.M., Ortin, F., Cueva, J.M.: Optimización de las primitivas de reflexión ofrecidas por los lenguajes dinámicos. In: Proceedings of the PROLE 2006, October 2006, pp. 53–64 (2006)
12. Rose, J.: Java specification request 292; supporting dynamically typed languages on the Java platform (2008), http://www.jcp.org/en/jsr/detail?id=292
13. Chiles, B.: Common language runtime inside out: IronPython and the Dynamic Language Runtime (2008), http://msdn2.microsoft.com/en-us/magazine/cc163344.aspx
14. OpenJDK: The Da Vinci machine (March 2008), http://openjdk.java.net/projects/mlvm/

15. Borning, A.H.: Classes versus prototypes in object-oriented languages. In: ACM/IEEE Fall Joint Computer Conference, pp. 36–40 (1986)
16. Lucas, C., Mens, K., Steyaert, P.: Typing dynamic inheritance: A trade-off between substitutability and extensibility. Technical Report vub-prog-tr-95-03, Vrije Un. Brussel (1995)
17. Pierce, B.P.: Types and Programming Languages. MIT Press, Cambridge (2002)
18. MetaSlash: PyChecker: a Python source code checking tool. Sourceforge (2008)
19. Meijer, E., Drayton, P.: Static typing where possible, dynamic typing when needed: The end of the cold war between programming languages. In: OOPSLA Workshop on Revival of Dynamic Languages (2004)
20. Ortin, F.: The StaDyn programming language (2008),
    http://www.reflection.uniovi.es
21. Chambers, C., Ungar, D.: Customization: Optimizing compiler technology for Self, a dynamically-typed oo programming language. In: ACM PLDI Conference (1989)
22. Udell, J.: D. languages and v. machines. Infoworld (August 2003)
23. Maes, P.: Computational Reflection. PhD thesis, Vrije Universiteit (1987)
24. Cazzola, W., Chiba, S., Saake, G.: Evolvable pattern implementations need generic aspects. In: ECOOP 2004 Workshop on Reflection, AOP, and Meta-Data for Software Evolution (2004)
25. Diehl, S., Hartel, P., Sestoft, P.: Abstract machines for programming language implementation. In: Future Generation Computer Systems, p. 739 (2000)
26. Meijer, E., Gough, J.: Technical overview of the CLR. Technical report, Microsoft (2000)
27. Singer, J.: JVM versus CLR: a comparative study. In: ACM Proceedings of the 2nd international conference on principles and practice of programming in Java (2003)
28. Stutz, D., Neward, T., Shilling, G.: Shared Source CLI Essentials. O'Reilly, Sebastopol (2003)
29. DeMichiel, L.G., Gabriel, R.P.: The common lisp object system: an overview. In: Bézivin, J., Hullot, J.-M., Lieberman, H., Cointe, P. (eds.) ECOOP 1987. LNCS, vol. 276, pp. 151–170. Springer, Heidelberg (1987)
30. Deutsch, L.P., Schiffman, L.A.: Efficient implementation of the Smalltalk-80 system. In: 11th annual ACM Symposium on Principles of Programming Languages, pp. 297–302 (1984)
31. Ancona, D., Anderson, C., Damiani, F., et al.: A type preserving translation of flickle into java. Electronic Notes in Theoretical Computer Science, vol. 62 (2002)
32. Serrano, M.: Wide classes. In: Guerraoui, R. (ed.) ECOOP 1999. LNCS, vol. 1628, p. 391. Springer, Heidelberg (1999)
33. Kleinder, J., Golm, G.: MetaJava: An efficient run-time meta architecture for Java. In: International Workshop on Object Orientation in Operating Systems, pp. 420–427 (1996)
34. Ungar, D., Chambers, G., Chang, B.W., Holzl, U.: Organizing programs without classes. In: Lisp and Symbolic Computation (1991)
35. Wolczko, M., Agesen, O., Ungar, D.: Towards a universal implementation substrate for object-oriented languages. Sun Microsystems Laboratories (1996)
36. CodeHaus: Boo. a wrist friendly language for the CLI (2008), http://boo.codehaus.org/
37. Bracha, G., Griswold, D.: Strongtalk: Typechecking Smalltalk in a production environment. In: OOPSLA 1993, ACM SIGPLAN Notices, vol. 28, pp. 215–230 (1993)
38. Ernst, E.: Dynamic inheritance in a statically typed language. Nordic Journal of Computing 6(1), 72–92 (1999)
39. Taivalsaari, A.: Kevo: A prototype-based OO language based on concatenation and module operations. Technical report, U. of Victoria, British Columbia (1992)
40. Project, T.: TheTunes project (March 2008), http://slate.tunes.org/
41. Cutsem, T.V., Mostinckx, S., Boix, E.G., Dedecker, J., Meuter, W.D.: AmbientTalk: Object-oriented event-driven programming in mobile ad hoc networks. In: XXVI International Conference of the Chilean Computer Science Society, SCCC 2007 (2007)

Scala Roles: Reusable Object Collaborations in a Library

Michael Pradel[1] and Martin Odersky[2]

[1] TU Dresden, Germany
michael@binaervarianz.de
[2] EPFL, Switzerland
martin.odersky@epfl.ch

Abstract. Purely class-based implementations of object-oriented software are often inappropriate for reuse. In contrast, the notion of objects playing roles in a collaboration has been proven to be a valuable reuse abstraction. However, existing solutions to enable role-based programming tend to require vast extensions of the underlying programming language, and thus, are difficult to use in every day work. We present a programming technique, based on dynamic proxies, that allows to augment an object's type at runtime while preserving strong static type safety. It enables role-based implementations that lead to more reuse and better separation of concerns.

1 Introduction

Software objects represent real-world entities. Often, those entities fundamentally change during their lifetime. Also, there may be different views on an object depending on the context. One way to express this in a programming language is by *dynamically adapting the type* of objects by enhancing and reducing the set of visible fields and methods.

Another issue tackled in this paper is *reuse*, one of the major goals in software engineering. Reuse is the process of creating software systems from existing software rather than building it from scratch. In object-oriented programming languages, classes abstract over sets of objects according to common properties and behavior. Though, relations between objects are usually manifold. In a traditional approach, this is reflected by a complex network of interconnected classes. As different contexts in which objects collaborate are not explicit in a purely class-based approach, reusing a particular object collaboration is difficult.

The software modeling community has been discussing since long a promising solution: *role modeling* or *collaboration-based design* [1,2,3]. The main idea is that objects play *roles*, each describing the state and behavior of an object in a certain context. In other words, a role provides a particular view on an object. Similar to objects, roles can be related to each other, for instance by references, field accesses, and method calls. A number of related roles form a *collaboration*. As objects can be abstracted by classes, roles are abstracted by *role types*, that can be thought of as partial classes. A typical example is an instance of a class Person that may play the role of a Professor (related to a role Student) and the role of a Father (related to Child and Mother).

J. Cordeiro et al. (Eds.): ICSOFT 2008, CCIS 47, pp. 23–36, 2009.

We can separate both concerns into two collaborations University and Family. Roles and collaborations permit to explicitly describe interwoven relations of objects, and hence, provide an interesting reuse unit orthogonal to classes.

While roles are generally accepted in modeling (see for example *UML collaborations* [4]), they are rare in today's programming languages. Most of the proposed solutions [5,6,7] either do not fully conform to the commonly used definition of the role concept [3] or require extensive changes in the underlying programming language. The latter makes them hard to employ in every day work since existing tools like compilers and debuggers cannot be used. In this work, we propose a lightweight realization of roles in the Scala programming language [8]. It can be realized as a library, that is, without any language extension.

Listing 1 gives a first glimpse of our solution. In the first line, a class Person is instantiated. The contexts in which the person occurs are represented by collaborations, that are instantiated in lines 4 and 5. In the following, the object is accessed playing the roles of a professor and a father (lines 8 and 11), as well as without any role (line 14). More details on our approach and other examples follow in Sections 3 and 4.

```
1   val p = new Person("John")
2
3   // collaborations are instantiated
4   val univ = new University{}
5   val fam = new Family{}
6
7   // person in the university context
8   (p as univ.professor).grade_students()
9
10  // person in the family context
11  (p as fam.father).change_diapers()
12
13  // person without any role
14  p.name    // "John"
```

Listing 1. A person playing different roles

The major thrust of this paper is to show that programming with roles is feasible in a lightweight fashion. More specifically, our contributions are the following:

- A programming technique for roles that might be applicable to most object-oriented programming languages. It is based on compound objects managed by dynamic proxies.
- A role library for Scala that allows to dynamically augment an object's type while preserving strong static type safety.
- A novel reuse unit, dynamic collaborations, that captures the relations of objects in a certain context.

The following section summarizes the idea of roles and collaborations in object-oriented programming languages. Our approach is explained in detail in Section 3, followed by Section 4 giving concrete examples of augmenting types and reusing the

Composite design pattern as a collaboration. Finally, a short overview of similar approaches is given in Section 5.

2   Objects and Roles

There is a wide range of different definitions of roles in the literature [9,10,1,2]. Steimann [3] gives a comprehensive overview and presents a list of role features, whereof the following are the most essential.

- A role comes with its own properties and behavior.
- Roles are dynamic, that is, they can be acquired and abandoned by objects at runtime. In particular, a role can be transferred from one object to another.
- An object can play more than one role at the same time. It is even possible to play two roles of the same role type multiple times (in different contexts).
- The state and the members of an object may be role-specific. That is, binding a role to an object can change its state as well as its fields and methods.

Roles are a relative concept, in the sense that a role never occurs alone, but always together with at least one other role. Related roles form a collaboration as shown in Figure 1, where a role drawn on top of an object indicates that the object plays the role. Possible relations between roles, such as references, field accesses, and method calls, are abstracted by a connecting line.

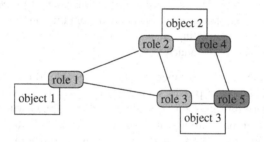

Fig. 1. Roles 1 to 3, as well as roles 4 and 5 form two distinct collaborations that can describe independent concerns

For a concrete example of the problem collaborations aim to solve, consider a graphics library containing a class `Figure` and two subclasses `BorderFigure` and `TextFigure` (Listing 2). These classes contain members representing properties like colors or the text in a `TextFigure`. Furthermore, we want to nest figures, for instance, putting a `TextFigure` inside a `BorderFigure`. This can be realized with the Composite pattern [11] which creates a tree-like data structure while allowing clients to treat individual objects and compositions of objects uniformly.

The code related to the Composite pattern is highlighted in Listing 2. We argue that, instead of being added to the figure classes, it should be extracted into a collaboration. This approach has two main advantages:

```
1  class Figure {
2    var bgColor = white
3
4    def addChild(f: Figure)
5    def removeChild(f: Figure)
6    def getParent: Figure
7    protected def setParent(f: Figure)
8  }
9
10 class BorderFigure extends Figure {
11   var borderColor = black
12
13   def addChild(f: Figure) = { /* ... */ }
14   // implementations of other abstract methods
15 }
16
17 class TextFigure extends Figure {
18   var text = ""
19
20   // implementations of abstract methods
21 }
```

Listing 2. A figure hierarchy implementing the Composite design pattern

- *Separation of concerns.* Figures have a number of inherent properties (in our example colors and text) that should be separated from the concern of nesting them.
- *Reuse.* Instead of implementing the pattern another time, we can reuse an existing implementation and simply attach it where needed.

Moving the highlighted code into supertypes is a reasonable solution in some cases. A role-based approach, however, provides more flexibility since behavior can be attached at runtime to arbitrary objects. Consequently, it can be applied without changing the type hierarchy of the graphics library, and even without access to their source code.

We propose to implement the Composite pattern as a collaboration with two roles parent and child. As a result, to access an instance of one of the figure classes being part of a composite, one can simply attach the parent or child role to it. As we will explain in more detail in the following sections, this is realized with the as operator in our approach. For instance, someFigure as composite.parent enhances the object someFigure with members related to being a parent in a composite.

3   Compound Objects with Dynamic Proxies

The purpose of this paper is to show how roles and collaborations can be realized in programming. This section explains our solution conceptually and shows details of our implementation in Scala that may be interesting for similar implementations in other languages. Our approach benefits from flexible language features of Scala, such as implicit conversions and dependent method types, and from the powerful mechanism of *dynamic proxies* in the Java API. The latter can be used since Scala is fully interoperable with Java. However, the described programming technique is not limited to a specific language. As we will argue at the end of this section, the essence of it may be carried over to other languages than Scala.

A major question is whether to implement roles with objects or as a first-class language construct. We opted for a lightweight solution that realizes roles as objects which are attached to core instances dynamically. One advantage is that the underlying programming language need not to be changed and existing tools like compilers can be used without modifications. However, this leads to the problem of having multiple objects where only one is expected by programmers. For instance, a person playing the role of a professor is represented by one core object of type `Person` and one role object of type `Professor`. Issues arising from that situation have been summarized as *object schizophrenia* [12]. The main problem to resolve is the unclear notion of object identity that can, for instance, lead to unexpected results when comparing objects.

We argue that one can avoid such confusion by regarding a core object with temporarily attached role objects as a *compound object*, as depicted in Figure 2. The compound object is represented to the outside by a dynamic proxy that delegates calls to the appropriate inner object.

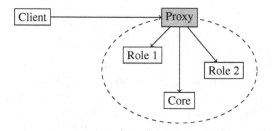

Fig. 2. The proxy intercepts calls to the compound object and delegates them via an invocation handler

A dynamic proxy is a particular object provided by the Java API. Its type can be set dynamically when creating it through a list of Java interfaces. Internally, dynamic proxies are realized by building and loading an appropriate class at runtime. The behavior of the proxy is specified reflectively by an *invocation handler*, that is, an object providing an `invoke` method that may delegate method calls to other objects.

We use a dynamic proxy to represent a role-playing object. Thus, its type is made up of the core object's type and the types of the role objects that are currently bound to it. The invocation handler of the proxy has a list of inner objects, one core object and arbitrary many role objects, and delegates calls to the responsible objects. Policies mapping method calls to inner objects may be specified if needed. A simple default is to reflectively delegate to role objects whenever they provide the required method and to the core object otherwise, such that roles may override the behavior of their core object.

Managing the compound object, creating a dynamic proxy with the appropriate type, and configuring the invocation handler is hidden from the user through a single operator called as. The expression object as role allows to access an object playing a certain role by temporarily binding role to object.

In Scala, all method calls can be written as infix operators. Hence, object as role is equivalent to object.as(role). However, we want to bind roles to arbitrary objects, that is, we cannot assume object to provide an as method. In contrast, we can easily add the inverse (but less intuitive to use) method playedBy to role objects. The trick is to use Scala's implicit conversions [13] to turn a method call object.as(role) into role.playedBy(object). An implicit conversion is a special method inserted by the compiler whenever an expression would not be otherwise type-correct.

```
1  implicit def anyRef2HasAs[Player <: AnyRef]
2      (core: Player) =
3    new HasAs[Player](core)
4
5  class HasAs[Player <: AnyRef]
6      (val core: Player) {
7    def as(r: Role[Player]) = r.playedBy(core)
8  }
```

Listing 3. An implicit conversion that adds the method as to arbitrary objects

The implicit conversion in lines 1 to 3 of Figure 3 wraps a core object into an instance of HasAs, a class providing the required as method. The method anyRef2-HasAs has a type parameter Player which is inferred from the type of the argument core. Player is restricted by the upper bound AnyRef, Scala's equivalent to Java's Object. The as method simply calls playedBy on the role object (line 7). This returns a proxy object having the type of the core object, Player, extended with the type of the role object, r.type. As a result, roles can be dynamically attached to arbitrary objects writing object as role, which gives type-safe access to core and role members.

An interesting detail to note here is that the return type of as is a *dependent method type*, an experimental feature of Scala. The return type of playedBy, and consequently also of the as method, is Player with r.type. Hence, the return type of the method depends on the value of its argument r.

3.1   Object Identity

Another issue is to provide a clear notion of *object identity*. We argue that the identity of an object should be the same independent of whether roles are attached or not. For instance, a person being a father is still the same person, and consequently, object identity should reflect this. To clarify the problem, consider comparing objects and role-playing objects. Four kinds of comparison are possible:

```
(1)                 object == (object as role)
(2)     (object as role) == object
(3)     (object as role) == (object as role)
(4)     (object as role1) == (object as role2)
```

To achieve (seeming) object equality between objects and role-playing objects, we modify identity-related methods of dynamic proxies. We use the fact that == and the equals method are equivalent in Scala. That is, the expressions x == y and x.equals(y) give the same result. We define equals and hashCode of proxies such that they map to the implementation of the core object, and, in case the right-hand operator of == is a proxy as well, compare with its core object. Although this solves the problem for expressions 2 to 4, it unfortunately does not for expression 1 since we cannot modify the equals and hashCode methods of arbitrary objects using a library approach. One possible solution would be to require the type of core objects to inherit from a type RolePlayer which contains an adapted equals methods. If the argument of equals is a proxy, it would compare with its core object, and otherwise fall back to the default implementation of equals. However, this makes adding roles to arbitrary objects impossible. Finding a satisfactory solution for this issue remains as future research.

So far, we have shown how arbitrary objects can be dynamically enhanced with new functionality by binding roles to them. Our solution provides strong static type safety, that is, only members of either the core object or a role object can be accessed without a type error. Moreover, obstacles arising from object schizophrenia can be solved with a compound object represented by a dynamic proxy and an adapted notion of object identity.

3.2 Collaborations with Nested Types

In the following, we delve into another important aspect of roles, namely grouping them into collaborations. Per definition, roles do not occur alone, but describe the behavior of an object in a certain context. This context is given by other roles yielding a set of related roles called collaboration.

One of the basic principles of Scala is nesting of types, allowing related types to be grouped and extended together by extending their outer type. Following this principle, a collaboration is presented by an outer trait[1] whose inner traits represent its roles. Our role library provides a trait Collaboration that concrete collaborations must extend. It contains an inner trait Role that must be extended to define concrete role types. Most of the details like the playedBy method creating the dynamic proxy, are implemented in these base traits, such that collaboration developers do not have to bother with it. Listing 4 shows a simplified version of the Collaboration trait.

At first, we create an invocation handler (line 4). It realizes the delegation of all incoming calls to either the core object or the role object and adds special treatment for the methods equals and hashCode. Then, createProxy is called. It reflectively retrieves Java interfaces for the core object and the role object (line 10) and instantiates a dynamic proxy via the Java API. The proxy has the type of the core object, Player, and of the role, this.type. Since newProxyInstance simply returns a java.lang.Object, we down-cast the proxy to Player with this.type before returning it to the user (line 5). This cast can be done safely as we configured the

---

[1] A trait is a type similar to a class, however, providing a safe form of multiple inheritance. Traits can be thought of as interfaces with an implementation. See [13] for further details.

```scala
1   trait Collaboration {
2     trait Role[Player <: AnyRef] {
3       def playedBy(core: Player): Player with this.type = {
4         val handler = new InvocationHandler(this, core)
5         createProxy(core, handler).asInstanceOf[Player with this.type]
6       }
7
8       private def createProxy(core: Player,
9                               handler: InvocationHandler) = {
10        val interfaces: Array[Class] = getInterfaces(this, core)
11        Proxy.newProxyInstance(core.getClass.getClassLoader,
12                               interfaces, handler)
13      }
14    }
15  }
```

Listing 4. The collaboration trait that concrete collaborations extend

proxy to have exactly this type. The benefit is that user code can be type-checked and no further casts are necessary.

In this section, we have shown how objects with dynamically changing types can be expressed with dynamic proxies. Furthermore, the flexibility of Scala permits to introduce a convenient syntax (the as operator), and hence, hide unnecessary details from the user. Collaborations can be expressed with nested types.

Although we implement our approach in Scala, we argue that its essence depends only on a small set of language features, and hence, can be transferred to other programming languages as well. There are two basic ingredients: First, one requires a way to dynamically create proxies whose type and implementation can be specified reflectively. Second, a notion of inner types is needed for collaborations. With minor adaptations, they can be realized with inner classes like those of Java. Other language features we used for our implementation, such as implicit conversions and dependent method types, help in providing a convenient syntax for using roles, but are not absolutely necessary.

4   Scala Roles in Action

The following section explains our approach from the perspective of collaboration developers and users. Collaboration-based programming has two major benefits. First, binding roles to objects and accessing them type-safely leads to a kind of *type dynamism*. It allows to enhance and reduce the set of visible members of objects at runtime without accessing its source code. Second, collaborations provide a *reuse unit* orthogonal to classes by encapsulating the behavior of multiple related objects. They focus on one specific aspect of a program, and thus, extract related code fragments. This leads to better separation of concerns and, possibly, reuse of a collaboration in different contexts. In this section we give concrete examples illustrating both benefits.

4.1 Persons and Their Roles

A classical example for roles are persons behaving specifically depending on the context, in other words, persons having different roles. Consider, for instance, the relation between a student at a university and his supervisor. The student gains motivation when being advised by the supervisor and wisdom when working, where the amount of gained wisdom depends on the student's current motivation. Assuming we have a class `Person` that may also occur in other contexts, supervisor and student can be modeled as roles of it. Listing 5 shows a collaboration with two roles `supervisor` and `student`, that are instances of the role types `Supervisor` and `Student`.

```
1   trait ThesisSupervision extends Collaboration {
2     val student = new Student{}
3     val professor = new Professor{}
4
5     trait Student extends Role[Person] {
6       var motivation = 50
7       var wisdom = 0
8       def work = wisdom += motivation/10
9     }
10    trait Supervisor extends Role[Person] {
11      def advise = student.motivation += 5
12      def grade =
13        if (student.wisdom > 80) "good"
14        else "bad"
15    }
16  }
```

Listing 5. A collaboration describing the relation between a student and a supervisor

A concrete collaboration must extend the abstract trait `Collaboration`. Doing so, it inherits the inner trait `Role` that can be extended by concrete roles. `Role` takes a type parameter that specifies the type of possible core objects playing the role. For roles that may be bound to arbitrary objects, `AnyRef` can be passed.

Listing 6 depicts how to use a collaboration. Paul supervises Jim's master project, and is, in his role as a PhD student, himself supervised by Peter, a professor. To use a collaboration, it must be instantiated (lines 8 and 9). Persons are accessed playing a certain role with the `as` operator. A role must be qualified with a collaboration instance. The main benefit of instantiating collaborations is that one object can play a role multiple times in different contexts.

Note that Paul plays different roles in the example. While he occurs as Jim's supervisor in line 12, he takes the role of a student in line 14. In each case, paul (seemingly) has a different type, and thus, offers different members. As our solution provides strong static type safety, calling the method `advise` on (`paul as phd.student`) would result in a type error during compilation.

It is also noteworthy that role-playing objects still have the type of their core object. For instance, in line 16, the field `name` that is defined in `Person` can still be accessed. Furthermore, a role itself can always access its current core object using a

```
1   // a master student
2   val jim = new Person("Jim")
3   // a PhD student
4   val paul = new Person("Paul")
5   // a professor
6   val peter = new Person("Peter")
7
8   val master = new ThesisSupervision{}
9   val phd = new ThesisSupervision{}
10
11  (jim as master.student).work
12  (paul as master.supervisor).advise
13
14  (paul as phd.student).work
15  (peter as phd.supervisor).grade
16  (peter as phd.supervisor).name
```

Listing 6. Usage of the `ThesisSupervision` collaboration. The person Paul plays different roles depending on the context.

method `core`. Hence, the implementation of the student role can, for example, access the student's name via `core.name`.

The state of the roles in a concrete collaboration instance is preserved between different uses of the `as` operator. These *stateful roles* allow transferring a role and its state from one core to another. For instance, suppose Peter retires as a professor at the end of Listing 6. A new professor can take over the supervision of Paul by binding the supervisor role to him: `newProf as phd.supervisor`.

4.2 Composite Design Pattern

The above example illustrates how roles enable runtime type enhancements. The second part of this section focuses on another benefit of roles and collaborations, namely reuse. In particular, we show how the Composite design pattern [11] can be represented as a reusable collaboration.

Expressed in terms of roles, the pattern essentially consists of a parent role and a child role [14]. Let us define a collaboration similar to that of Listing 5 with two roles `parent` and `child` providing the functionality of a composite, that is, methods `addChild`, `removeChild`, etc. Having exactly one instance of each role in the collaboration would imply dealing with a new collaboration instance for each parent-child relation between two objects. As a more convenient solution, we propose *role mappers*, helper objects creating a new role instance whenever a new core object requires a certain role. Before delving into details of its implementation, Listing 7 shows how to use the Composite collaboration.

In contrast to Listing 2, the figure classes do not contain any source code for figures being a composite (lines 1 to 9). Instead, we instantiate the Composite collaboration in line 16 and parametrize it with the desired type of core objects. To enhance the readability of the following code, two implicit conversions are defined in lines 17 and 19. They cause figures to be converted into figures playing either the parent role or the child role. Hence, the figures can be used as if they contained composite members, such as addChild. Alternatively, we could also use the as operator explicitly, for instance, writing (f4 as c.child).getParent in line 27.

```
1   class Figure {
2     var bgColor = white
3   }
4   class BorderFigure extends Figure {
5     var borderColor = black
6   }
7   class TextFigure extends Figure {
8     var text = ""
9   }
10
11  val f1 = new Figure
12  val f2 = new TextFigure
13  val f3 = new BorderFigure
14  val f4 = new TextFigure
15
16  val c = new Composite[Figure]{}
17  implicit def figure2parent(f: Figure) =
18    f as c.parent
19  implicit def figure2child(f: Figure) =
20    f as c.child
21
22  f1.addChild(f2)
23  f1.addChild(f3)
24  f3.addChild(f4)
25
26  f1.getChild(0) // f2
27  f4.getParent   // f3
```

Listing 7. With the Composite collaboration, figures can be treated as members of a composite without containing the implementation of the pattern

Contrary to the thesis supervision example, there exist an arbitrary number of instances of each role type in the Composite collaboration. Instead of instantiating roles statically as in Listing 5, there are two role mappers parent and child. A role mapper deals with the binding between core objects and role instances, creating new role instances on demand and reusing existing ones for already known core objects. Similarly to a role, a role mapper provides a playedBy method. Hence, using the same syntax as for roles with a fixed number of instances per collaboration, role mappers allows for arbitrary many of them. Whether to use roles or role mappers is a design decision of collaboration developers.

5 Related Work

There are a couple of other interesting approaches towards implementing roles. The Role Object pattern [5] describes a design that splits one conceptual object into a core object and multiple role objects, each enhancing the core object for a different context. Core classes and role classes extend a common superclass that clients deal with. Clients add (remove) roles to (from) a core object by calling appropriate methods on it and passing a role descriptor as argument. We developed the ideas of the Role Object pattern focusing on two major drawbacks: First, clients can only dynamically detect if a core object provides a certain role and if so, must down-cast the role object before invoking role-specific methods. Hence, instead of having static type safety, programmers need to deal with runtime checks. Second, the role object pattern suffers from the problem of object schizophrenia; thus, clients must pay attention to not rely on object identity.

Steimann proposes two independent type hierarchies, one for classes (called *natural types*) and one for role types, and presents a role-oriented modeling language formalizing his approach [3]. We do not adopt this idea mainly for pragmatic reasons, since its realization in an existing programming language demands substantial changes to the type system. Also, two strictly separated type hierarchies contradict one of the properties of roles in [3], namely roles playing roles.

A recent and very inspiring work on roles is ObjectTeams/Java [6]. This is an extension of Java adding first-class support for roles, role types, and collaborations (called *teams*). Collaborations are represented as special classes, *team classes*, whose inner classes are considered to be role types. The problem that role objects do not directly conform to the type of their core objects is solved by *translation polymorphism*, an implicit type-safe conversion between role instances and their core instances [15].

Roles can also be realized with aspect-oriented programming [16]. Kendall analyzes how to implement role models with aspects and compares it with object-oriented approaches [17]. In [18], design patterns are implemented with AspectJ, leading to similar benefits as our approach, namely reusability and better separation of concerns.

Finally, this paper also relates to work on first-class relationships [19,20,21], where the way objects relate to each other is described through appropriate language constructs. We show in [22] how the Scala Roles library can be used to achieve similar goals by making relationships explicit, however, without requiring new language constructs.

6 Conclusions

In this paper we propose a programming technique that enables expressing roles and collaborations. It is lightweight in the sense that no changes to the underlying language are necessary. Instead, dynamic proxies solve the problem of representing multiple objects as one compound object. Our approach allows to widen the set of members of an object at runtime, or in other words, to dynamically augment its type. Nevertheless, user code can be statically type-checked. We provide a Scala library as proof-of-concept and show how extracting concerns into collaborations supports reuse.

Future work will include applying our approach to a larger project in order to verify its usefulness and gain more insight about roles in programming languages. Another open task is a role-based library of reusable implementations of design patterns and other recurring object collaborations. Moreover, other role features should be investigated further, such as roles restricting access to its core object and restrictions on the sequence in which roles may be acquired and relinquished.

Acknowledgements. We would like to thank Prof. Uwe Aßmann for inspiring this work and the anonymous reviewers for their comments and suggestions.

References

1. Reenskaug, T., Wold, P., Lehne, O.A.: Working with Objects, The OOram Software Engineering Method. Manning (1996)
2. Riehle, D.: Framework Design: A Role Modeling Approach. PhD thesis, ETH Zürich (2000)
3. Steimann, F.: On the representation of roles in object-oriented and conceptual modelling. Data & Knowledge Engineering 35(1), 83–106 (2000)
4. Object Management Group OMG: OMG Unified Modeling Language (OMG UML), Superstructure, v2.1.2 (November 2007)
5. Bäumer, D., Riehle, D., Siberski, W., Wulf, M.: Role Object. In: Pattern Languages of Program Design 4, pp. 15–32. Addison-Wesley, Reading (2000)
6. Herrmann, S.: A precise model for contextual roles: The programming language ObjectTeams/Java. Applied Ontology 2(2), 181–207 (2007)
7. Smaragdakis, Y., Batory, D.: Mixin layers: An object-oriented implementation technique for refinements and collaboration-based designs. ACM Transactions on Software Engineering and Methodology (TOSEM) 11(2), 215–255 (2002)
8. Odersky, M.: Scala Language Specification, Version 2.7 (May 2008)
9. Guarino, N.: Concepts, attributes and arbitrary relations: Some linguistic and ontological criteria for structuring knowledge bases. Data & Knowledge Engineering 8(3), 249–261 (1992)
10. Kristensen, B.B., Østerbye, K.: Roles: Conceptual abstraction theory and practical language issues. Theory and Practice of Object Systems 2(3), 143–160 (1996)
11. Gamma, E., Helm, R., Johnson, R., Vlissides, J.: Design Patterns: Elements of Reusable Object-Oriented Software. Addison-Wesley, Reading (1995)
12. Harrison, W.: Homepage on subject-oriented programming and design patterns (1997), http://www.research.ibm.com/sop/sopcpats.htm
13. Odersky, M., Spoon, L., Venners, B.: Programming in Scala, A comprehensive step-by-step guide. Artima (2008)
14. Riehle, D.: Composite design patterns. In: Conference on Object-Oriented Programming, Systems, Languages, and Applications (OOPSLA 1997), pp. 218–228. ACM, New York (1997)
15. Herrmann, S., Hundt, C., Mehner, K.: Translation polymorphism in Object Teams. Technical Report 2004/05, TU Berlin (2004)
16. Kiczales, G., Lamping, J., Mendhekar, A., Maeda, C., Lopes, C.V., Loingtier, J.M., Irwin, J.: Aspect-oriented programming. In: Aksit, M., Matsuoka, S. (eds.) ECOOP 1997. LNCS, vol. 1241, pp. 220–242. Springer, Heidelberg (1997)
17. Kendall, E.A.: Role model designs and implementations with aspect-oriented programming. SIGPLAN Notices 34(10), 353–369 (1999)
18. Hannemann, J., Kiczales, G.: Design pattern implementation in Java and AspectJ. SIGPLAN Notices 37(11), 161–173 (2002)

19. Rumbaugh, J.E.: Relations as semantic constructs in an object-oriented language. In: Conference on Object-Oriented Programming, Systems, Languages, and Applications (OOPSLA 1987), pp. 466–481 (1987)
20. Bierman, G.M., Wren, A.: First-class relationships in an object-oriented language. In: Black, A.P. (ed.) ECOOP 2005. LNCS, vol. 3586, pp. 262–286. Springer, Heidelberg (2005)
21. Balzer, S., Gross, T.R., Eugster, P.: A relational model of object collaborations and its use in reasoning about relationships. In: Ernst, E. (ed.) ECOOP 2007. LNCS, vol. 4609, pp. 323–346. Springer, Heidelberg (2007)
22. Pradel, M.: Explicit relations with roles - a library approach. In: Workshop on Relationships and Associations in Object-Oriented Languages (RAOOL) at OOPSLA 2008 (2008)

Common Criteria Based Security Scenario Verification

Atsushi Ohnishi

Department of Computer Science, Ritsumeikan University, Japan
ohnishi@cs.ritsumei.ac.jp

Abstract. Software is required to comply with the laws and standards of software security. However, stakeholders with less concern regarding security can neither describe the behaviour of the system with regard to security nor validate the system's behaviour when the security function conflicts with usability. Scenarios or use-case specifications are common in requirements elicitation and are useful to analyze the usability of the system from a behavioural point of view. In this paper, the authors propose both (1) a scenario language based on a simple case grammar and (2) a method to verify a scenario with rules based on security evaluation criteria.

1 Introduction

Scenarios are important in software development [5], particularly in requirements engineering [1], since they provide concrete system description [15,18]. Moreover scenarios are useful in defining system behaviors done by system developers and validating the requirements undertaken altogether with customers [4]. In many cases, scenarios become foundation of system development. Incorrect scenarios will lead to negative impact on system development process in overall. However, scenarios are informal and it is difficult to verify the correctness of scenarios. The errors in incorrect scenarios may include:

1. Vague representations,
2. Lack of necessary events,
3. Extra events,
4. Wrong sequence among events.

The authors have developed a scenario language for describing scenarios in which simple action traces are embellished to include typed frames based on a simple case grammar [6] of actions and to describe the sequence among events [20]. Since this language is a controlled language, the vagueness of the scenario written using this language can be reduced [11]. Furthermore, the scenario created with this language can be transformed into internal representation. In the transformation, both lack of cases and illegal usage of noun types can be detected, and concrete words will be assigned to pronouns and omitted indispensable cases [11]. As a result, the scenario with this language can avoid errors typed 1 previously mentioned.

Furthermore, software security requirements affect the whole behavior of the software system and not only parts of the system. Most stakeholders may not be software

J. Cordeiro et al. (Eds.): ICSOFT 2008, CCIS 47, pp. 37–47, 2009.

security professionals. Almost all users and clients of the system will have no knowledge about software security. However, they still feel that it is important to comply with the laws and standards for information systems and software security.

Therefore, they may suggest requirements to comply with such standards although they may not be able to envision the behavior of the system once these suggestions are incorporated. Consequently, it is necessary for them to leave validation of requirements about their business rules to developers who do not have knowledge about the business rules.

Although developers have knowledge about software security, this is usually limited to general knowledge. They cannot decide who to apply the techniques of software security to specific business rules. If the system satisfies the laws and the standards, the users may find that the behavior differs from that initially envisioned after the completion of development.

Behaviors related to software security often conflict with other requirements, such as usability, cost and performance. Therefore, the developers cannot include all functional and non-functional requirements regarding software security in the software requirements specification.

There are a number of reasons why it is necessary to focus on elicitation of security and usability requirements. The requirements regarding usability will conflict with security requirements. The remaining requirements, such as cost or performance requirements, can be resolved by increasing other resources.

We focus on the verification method of scenarios with rules [17] and customize the rules to satisfy the software security common criteria [8].

2 Scenario Language

2.1 Outline

Our scenario language has been already introduced in several papers such as in [20,11]. However, in this paper, a brief description of this language will be given for convenience.

A scenario can be regarded as a sequence of events. Events are behaviors employed by users or systems for accomplishing their goals. We assume that each event has just one verb, and that each verb has its own case structure. The scenario language has been developed based on this concept. Verbs and their own case structures depend on problem domains such as elevator control [11], PC chair's job [3] and train ticket reservation [12], but the roles of cases are independent of problem domains. The roles include agent, object, recipient, instrument, and source, etc [6,11].

We provide Requirements Frames [11] in which verbs and their own case structures are specified. This frame depends on problem domains. Each action has its case structure, and each event can be transformed into internal representation based on this frame. In the transformation, concrete words will be assigned to pronouns and omitted indispensable cases. With Requirements Frame, we can detect both the lack of cases and the illegal usage of noun types [11].

We assume four kinds of time sequences among events:

1) sequential,
2) selective,
3) iterative, and
4) parallel.

Actually most events are sequential events. Our scenario language defines the semantic of verbs with their case structure. For example, data flow verb has source, goal, agent, and instrument cases. Since such case structure can define the abstraction level, scenario provided using our scenario language becomes the almost same level of the abstraction.

2.2 Scenario Example

We consider a scenario of train ticket reservation in a railway company. Figure 1 shows a scenario of customer's purchasing a ticket of express train at a service center. This scenario is written with our scenario language based on a video that records behaviors of both a user and a staff at one particular service center.

A title of this scenario is given at the first two lines in Fig.1. Viewpoints of considered scenario are specified at the third line. In this paper, viewpoints mean active objects such as human or system appearing in the scenario. There exist two viewpoints, namely staff and customer. The order of specified viewpoints means the priority. In this example, the first featured object is staff and the second one is customer. In such a case, the former becomes the subject of an event.

In addition, pre-condition specifies a condition that satisfies at the start of the scenario. Post-condition specifies a condition that satisfies at the end of the scenario.

```
[Title: A customer purchases a train ticket of reservation seat]
[Viewpoints: Staff, customer]
[Pre-condition: the customer has enough money to buy a ticket &
has not a ticket & has not reserved a seat]
[Post-condition: the customer will get a ticket & reserved a seat]
{
1. A staff asks a customer about leaving station and destination as customer's request.
2. He sends the customer's request to reservation center via private line.
3. He retrieves available trains with the request.
4. He informs the customer of a list of available trains.
5. The customer selects a train that he/she will get.
6. The staff retrieves available seats of the train.
7. He shows a list of available seats of the train.
8. The customer selects a seat of the train.
9. If (there exists a seat selected by the customer) then the staff reserves the seat with the terminal.
10. He gets a permission to issue a ticket of the seat from the center.
11. The customer paid for the ticket by cash.
12. He gives the ticket to the customer.
}
```

Fig. 1. Scenario example

In this scenario, most events are sequential, except one selective event (the 9th event). Selection can be expressed with if-then syntax like program languages. Actually, event number is for reader's convenience and not necessary.

3   Verification of Scenarios

When a scenario is described, necessary events may be missing, unnecessary events may be mixed or time sequence among events may be inaccurate. These errors may have a negative impact on system development; therefore, it is necessary to detect these errors. The errors, also employed as correctness verification items, of scenarios include:

1. Lack of necessary events
2. Extra events
3. Wrong time sequence among events

We can check whether an event is lacking, being extra one or being sufficient one by comparing its correct occurrence times with the times that it occurred in the scenario. Similarly, we can check whether the time sequence among events is wrong by comparing the correct time sequence with the time sequence described in the scenario.

We propose a method to verify the correctness of scenarios by using rules to detect the errors in scenarios. We assume that a rule is a description of the correct occurrence times of an event and/or the correct time sequence among events, which the scenario ought to satisfy. One scenario may be verified with several rules.

3.1   Rule

Rule is composed of the description of rule's event and the description of event's occurrence times and/or time sequence among events. In this sense, our rules just specify the occurrence of events and/or the sequence of events. If the abstraction level of events of rules becomes high, the rules can be applied to scenarios of several different domains.

Events in Rule. In a rule, there are one or more events, whose occurrence times and/or time sequences are specified. When a scenario is verified with a rule, the rule's events can correspond to the scenario's events. By finding the corresponding events in the scenario, and by checking the occurrence times and/or the time sequence of these events, one scenario can be verified.

As a result, it is necessary to get the corresponding relation between the rule's events and the scenario's events. For this reason, the rule's events are also described based on Requirements Frames, and can also be transformed into the internal representation. If the rule's event and the scenario's event have the same internal representation, then they are deemed as corresponding events.

In order to improve the verification effect, it is not sufficient that the corresponding relation between the rule's event and the scenario's event is 1 to 1 ratio. It is expected that a rule's event can correspond to several scenario's events. As a result, the occurrence times and/or the time sequence of these scenario's events can be checked with one rule. In such a case, the rule's event has an abstract representation, and the corresponding scenario's events have several concrete representations.

For the above reason, we permit the abstract description of rule's event. An abstract event may be transformed into several concrete events, when finding its corresponding events in the scenario. At this time, the corresponding relation between the rule's event and the scenario's event is 1 to many. There are two kinds of abstract events in the rule.

1. Some indispensable cases are omitted in the event sentence.
2. There include "something," "someone," "same thing," "same one," etc. in the event sentence.

In the first kind of abstract events, the omitted cases fit any noun. This kind of abstract events will be transformed into concrete events by substituting concrete nouns for omitted cases, when finding its corresponding events in observed scenario. For example, a rule's event "system feedbacks to user" can be transformed into an internal representation.

This event can correspond to any scenario's event whose action is "feedback, etc.", agent case is "system", and recipient case is "user". In the second kind of abstract events, "something" / "someone" is similar to the omitted case in the first kind of abstract events, and fit anything / anyone. The reason of dividing abstract events into two kinds should be explained. We assume that there is a rule that describes the time sequence among events. Under this rule there exist two separate events and under these two events there exist a case A and a case B. We want to specify that case A and case B can have any content but they have to be the same content. If we simply omit case A and case B in the rule description, it cannot be warranted that case A and case B have the same content. By specifying "something" / "someone" for case A, "same thing" / "same one" for case B, case A and case B can have any content, and they are the same content.

In the second kind of abstract events, "same thing" / "same one" fits the same noun with "something" / "someone" that appears in the same rule. This kind of abstract events will be transformed into concrete events by substituting concrete nouns for "something" / "someone" / "same thing" / "same one", etc. when finding its corresponding events in the scenario.

Occurrence Times of an Event. The correct occurrence times of an event, which the scenario ought to satisfy, can be specified as a rule. By comparing the correct occurrence times with the times that this event occurred in the scenario, whether this event is lack of or excess of occurrence can be checked. The occurrence times of an event are described based on regular expression as follows.

1. E: event E occurs just one time.
2. E+: event E occurs one or more times.
3. E?: event E occurs one time or does not occur.
4. E!: event E never occurs.
5. E{m}: event E occurs m times.
6. E{m,}: event E occurs m or more times.
7. E{,n}: event E occurs n or less times.
8. E{m,n}: the occurrence times of event E is from m to n.

We adopted the syntax of regular expression and similar its semantics.

Time Sequence among Events. The correct time sequence among events that scenario ought to satisfy can be specified as rules. By comparing the correct time sequence with the time sequence described in the scenario, time sequence among events can be checked. According to the time sequence in the scenario described in section 2.1, we assume the following rules.

1. *Before/After E1, E2: Before/After event E1 occurs, event E2 should occur.*
2. *If (condition) (E1, E2): Event E1 and event E2 occur selectively. If the condition is true, event E1 occurs. If the condition is false, event E2 occurs.*
3. *Do (E1,E2,...) until(condition): Until the condition becomes true, event E1, E2, ... occur iteratively.*
4. *AND(E1,E2,...): All of the events E1, E2, and others parallel occur.*
5. *OR(E1,E2,...): One of the events E1, E2, ... or more parallel ccurs.*
6. *XOR(E1,E2,...): Just one of the events E1, E2, ... occurs.*

As previously described, when a scenario is verified with a rule that includes the abstract event, it is possible that an abstract event corresponds to several scenarios' events. In this case, it is necessary to check the time sequence of every corresponding event in the scenario and the results should be shown one by one.

Scenario-checking with Rule. Our scenario-checking procedure consists of two phases. The first phase is selection of applicable rules from rule DB. We specify both (1) pre-conditions and post-conditions and (2) viewpoints in each of the rules. When the conditions and viewpoints of a rule are much the same as those of a given scenario, the rule is selected for checking the scenario.

The second phase is analysis of rules and checking the consistency between a rule and the scenario as described in 3.1. The result will be passed to a checking-system user. The scenario checking can be achieved by automatically checking whether the scenario satisfies the rules, through the internal representation of scenario and the internal representations of rules.

We firstly find events in a scenario each of which corresponds to an event in a rule as described in 3.1.1. When a scenario is checked with a rule, the occurrence times and/or the time sequence of corresponding events in the scenario will be checked. We provide two checking procedures. One is for checking the occurrence time and the other is for checking the time sequence. Figure 2 shows outline of checking procedure for the occurrence time of an event and Figure 3 shows outline of checking procedure for the time sequence between events E1 and E2.

3.2   Evaluation

We have developed a prototype system based on the method. There exist 35 errors in 15 scenarios. These 35 errors can be classified into three categories, namely (1) wrong sequence of events, (2) lack of events, and (3) extra events. The number of errors grouped into the above categories are 16, 8 and 11, respectively. We could detect part of these errors with our method shown in Table 1.

The detection ratio is 63%. The reason why our method seems to be weak for detecting extra events is that it is very difficult to predict extra events and make rules for

Initialize a counter (counter=0).
Find the scenario's event that corresponds to the rule's event
while (the corresponding event in the scenario will be found)
do
counter=counter+1
Show the corresponding event and its occurrence condition to user.
Find the next scenario's event that corresponds to the rule's event.
od
Compare the occurrence times specified in the rule with the counter, and show the result.

Fig. 2. Checking procedure of the occurrence of events

Find the scenario's event that corresponds to E1 from the beginning of the scenario.
if (the corresponding event of E1 not be found)
then show this error, and the checking ends.
else do
Show the corresponding event and its occurrence condition.
Find an event that corresponds to E2 and satisfies the time sequence
if (the corresponding event of E2 not be found)
then show the result that the scenario does not satisfy the rule
else do
Show the corresponding event and its occurrence condition.
Find the next scenario's event that corresponds to E2 and satisfies the time sequence.
until (the corresponding event of E2 not be found)
fi
Find the next scenario's event that corresponds to E1.
until (the corresponding event of E1 not be found)
fi

Fig. 3. Checking procedure of the sequence of events

them in advance. On contrast, it is not so difficult to predict indispensable events and correct sequence of events and make rules for them in advance. Another reason why our method is strong for detecting wrong sequence of events and lack of events is that rules for the occurrence time of events are effective to detect lack of events and rules for the sequence of events are effective to detect wrong sequence of events. To improve the detection ratio, we have to introduce another type of rules for detecting extra events.

The describers can easily correct the detected errors. Since scenario writers can determine the abstraction level of scenarios, the number of events may differ depending on the scenario writers. Causes of undetected errors related to both lack of events and extra events are misunderstandings of the reservation jobs.

3.3   Rule Based on Common Criteria

The security evaluation criteria suite used in this paper is ISO/IEC 15408 Evaluation Criteria for IT Security [8]. The evaluation criteria suite is useful to verify scenarios because some of them are easily represented as rules. We can detect scenarios which do not satisfy rules based on the security evaluation criteria.

Table 1. Detected errors in scenario

	The number of errors	The number of detected errors
Wrong sequence of events	16	14
Lack of events	8	6
Extra events	11	2
Total	35	22

7. Class FIA: Families of Identification and authentication

7.1 Authentication failures (FIA_AFL) require that the system be able to terminate the session establishment process after a specified number of unsuccessful user authentication attempts. It also requires that, after termination of the session establishment process, the system be able to disable the user account or the point of entry from which the attempts were made until an administrator-defined condition occurs.

FIA_AFL.1.1.
The security function shall detect when [selection: [assignment: positive integer number], an administrator configurable positive integer within [assignment: range of acceptable values]] unsuccessful authentication attempts occur related to [assignment: list of authentication events].

FIA_AFL.1.2.
When the defined number of unsuccessful authentication attempts has been met or surpassed, the security function shall [assignment: list of actions].

Fig. 4. An excerpt of security evaluation criteria

```
[Authentication failures][system, user]
{
The system requests the user for a password.
The system receives the password from the user.
The system authenticates the user via the password
if ( unsuccessful authentication attempts meet or exceeds 3 times ) then
The system switches to "Invalidate ID".
fi
}
```

Fig. 5. An example of a scenario

As for FIA_AFL.1.1 and 1.2, we can provide a rule that authentication event should occur n or less times, where n is an acceptable positive integer and another rule that authentication event occurs n times or more, event about security function should occur.

We can verify the scenario shown in Fig. 5 with rules shown in Fig. 4 and confirm the scenario satisfies the rules. If the authentication event occurs more than three times, the verifier can detect the unsuccessful status and also detect the switching to "invalidate ID." These results will be provided to a user and he will judge the correctness of the scenario.

> This family defines the types of user authentication mechanisms supported by the TSF (Target of evaluation Security Functions). This family also defines the required attributes on which the user authentication mechanisms must be based.
>
> FIA_UAU.1.1
> The TSF shall allow [assignment: list of TSF mediated actions] on behalf of the user to be performed before the user is authenticated.

Fig. 6. Sequential events in user autentication

Another example is FIA_UAU (User Authentication).1.1 shown below.

We can provide a rule that events of TSF mediated actions occur before event of user authentication.

Based on the proposed method, we have developed a prototype system on a Windows XP PC with java program language.

4 Discussion

ISO/IEC 15408 is a most commonly used security evaluation criteria suite and it has descriptions of functional requirements classified according to purpose and function of IT system to be developed. The suite also has descriptions of simple behaviors to meet the functional requirements.

In the security evaluation criteria suite, there are 68 families of functional requirements into 11 components, such as encryption, authentication, etc. In this paper, we focus mainly on usability, and consider 12 functional requirements families and other related families, because these families can be easily represented as rules for the verification.

The remaining requirements families are for quality of function or evaluation of the functions themselves, and so they do not affect the behaviors that can be seen by the users. The remaining families are difficult to transform rules. This point is a major problem of the proposed research. Actually, we can represent 23 rules for security requirements of the security evaluation criteria, while 44 security requirements of the criteria cannot be represented as rules.

In this paper, we consider "7.1 Authentication failures" and show the verification process of the scenario with rules.

The rules must be written by security professionals to detect wrong scenarios which involve incorrect security behavior that causes a critical vulnerability in the system. The professionals can rewrite the scenario to meet the security evaluation criteria into scenario easily.

Some criteria are not suitable to represent as rules and scenarios cannot be verified with these points of view. To solve this problem is left as a future work.

5 Related Works

Araujo and Whittle et al. proposed an analysis method in their scenario description process [2,19]. This method focuses on generation of state machines by synthesizing scenarios and validating their correctness. This method is very useful for requirements analysis and the design process after requirements elicitation. Instead, our method focuses on validation in the requirements elicitation phase and finding the conflicts of requirements.

Sindre and Opdahl proposed Misuse Cases [14] and McDermott and Fox proposed Abuse Cases [9] for security requirements elicitation. These methods are useful for brainstorming or discussion by clarifying the threats. However, our method uses the security evaluation criteria and focuses on the comprehensive elicitation of security requirements.

SIREN is a security requirements management method that focuses on security evaluation criteria or common criteria [16], but this method focuses on reuse of requirements specifications regarding security requirements and they did not mention the behavior of the security requirements functions.

Sutcliffe et al. propose a verification method of scenarios based on validation-frame [15]. This frame consists of situation part and requirements part. In situation part, pattern of events and actions are defined. In requirements one, some generic requirements are needed to handle each of these patterns. Using validation-frame, crosschecking between scenario and requirements is possible. Our approach is similar, but we enable to check (1) wrong sequence of events and (2) the number of occurrence of events. On contrast, validation-frame does not check them.

6   Conclusions

The author proposed a scenario checking method with rules based on the security evaluation criteria. We can specify the occurrence times of events and/or the time sequence among events as rules. Both scenario and rules can be transformed into the internal representation so that we can check scenario with rules and evaluate the correctness of one particular observed scenario.

The proposed method was demonstrated by the example and was evaluated. The evaluation results show that errors (the lack of events, extra events, the wrong sequence among events, and wrong behaviors against the security common criteria) in scenario can be effectively detected by checking the scenario with rules. By using this correctness checking method, we can get a scenario that satisfies security common criteria more effectively in system development.

Acknowledgments. The author thanks to Dr. Hiroya Itoga, Mr. Tatsuya Toyama and Mr. Kenta Nishiyuki for their contributions to the research.

References

1. Alexander, I.F., Maiden, N.: Scenarios, Stories, Use Cases - Through the Systems Development Life-Cycle. John Wiley & Sons, Chichester (2004)
2. Araujo, J., Whittle, J., Kim, D.: Modeling and Composing Scenario-Based Requirements with Aspects. In: 12th International Requirements Engineering Conference (RE 2004), pp. 58–67 (2004)

3. Barish, R.: ACM Conference Committee Job Description, Conference Manual, Section No. 6.1.1 (1997),
   `http://www.acm.org/sig_volunteer_info/conference_manual/6-1-1PC.HTM`
4. Carroll, J.M.: Making Use: Scenario-based Design of Human Computer Interactions. MIT Press, Cambridge (2000)
5. Cockburn, A.: Writing Effective Use Cases. Addison Wesley, USA (2001)
6. Fillmore, C.J.: The Case for Case, Universals in Linguistic Theory. Bach & Harms, Holt, Rinehart and Winston Publishing, Chicago (1968)
7. IEEE Std. 830-1998, IEEE Recommended Practice for Software Requirements Specifications (1998)
8. ISO/IEC 15408 common criteria (2005)
9. McDermott, J., Fox, C.: Using Abuse Case Models for Security Requirements Analysis. In: Proceedings of the 15th IEEE Annual Computer Security Applications Conference (ACSAC 1999), pp. 55–65 (1999)
10. Ohnishi, A.: Software requirements specification database based on requirements frame model. In: Proceedings of the Second IEEE International Conference on Requirements Engineering (ICRE 1996), pp. 221–228 (1996)
11. Ohnishi, A., Potts, C.: Grounding Scenarios in Frame-Based Action Semantics. In: Proc. of 7th International Workshop on Requirements Engineering: Foundation of Software Quality (REFSQ 2001), June 4-5, pp. 177–182. Interlaken, Switzerland (2001)
12. Railway Information System Co., Ltd., JR System (2001),
   `http://www.jrs.co.jp/keiki/en/index_main.html`
13. Schneier, B.: Secrets & Lies Digital Security in a Networked World. John Wiley & Sons, Chichester (2001)
14. Sindre, G., Opdahl, A.L.: Eliciting security requirements with misuse cases. Requirements Engineering 10, 34–44 (2005)
15. Sutcliffe, A.G., Maiden, N.A.M., Minocha, S., Manuel, D.: Supporting Scenario-Based Requirements Engineering. IEEE Trans. Software Engineering 24(12), 1072–1088 (1998)
16. Toval, A., Nicolaus, J., Moros, B., Gracia, F.: Requirements Reuse for Improving Information Systems Security: A Practitioner's Approach. Requirements Engineering 6(4), 205–219 (2002)
17. Toyama, T., Ohnishi, A.: Rule-based Verification of Scenarios with Pre-conditions and Post-conditions. In: Proc. Of the 13th IEEE International Conference on Requirements Engineering (RE 2005), Paris, France, pp. 319–328 (2005)
18. Weidenhaupt, K., Pohl, K., Jarke, M., Haumer, P.: Scenarios in System Development: Current Practice. IEEE Software 15(2), 34–45 (1998)
19. Whittle, J., Araujo, J.: Scenario modeling with aspects. IEE Proceedings Software Special Issue 151(4), 157–172 (2004)
20. Zhang, H., Ohnishi, A.: Transformation between Scenarios from Different Viewpoints. IEICE Transactions on Information and Systems E87-D(4), 801–810 (2004)

A Software Infrastructure for User–Guided Quality–of–Service Tradeoffs

João Pedro Sousa[1], Rajesh Krishna Balan[2], Vahe Poladian[3], David Garlan[3], and Mahadev Satyanarayanan[3]

[1] Computer Science Department, George Mason University
4400 University Drive, Fairfax VA, USA
[2] School of information Systems, Singapore Management University
80 Stamford Road, Singapore
[3] Computer Science Department, Carnegie Mellon University
5000 Forbes Avenue, Pittsburgh PA, USA
jpsousa@cs.gmu.edu, rajesh@smu.edu.sg
{poladian,garlan,satya}@cs.cmu.edu

Abstract. This paper presents a framework for engineering resource-adaptive software targeted at small mobile devices. Rather than building a solution from scratch, we extend and integrate existing work on software infrastructures for ubiquitous computing, and on resource-adaptive applications.

This paper addresses two research questions: first, is it feasibility to coordinate resource allocation and adaptation policies among several applications in a way that is both effective and efficient. And second, can end-users understand and control such adaptive behaviors dynamically, depending on user-defined goals for each activity. The evaluation covered both the systems and the usability perspectives, the latter by means of a user study.

The contributions of this work are: first, a set of design guidelines, including APIs for integrating new applications; second, a concrete infrastructure that implements the guidelines. And third, a way to model quality of service tradeoffs based on utility theory, which our research indicates end-users with diverse backgrounds are able to leverage for guiding the adaptive behaviors towards activity-specific quality goals.

Keywords: Mobile computing, Resource adaptation, Self-adaptive systems, Software architecture, User studies.

1 Introduction

Sophisticated software is increasingly being deployed on small mobile devices, taking advantage of their growing capabilities and popularity. Media streaming is already found frequently in PDAs and high-end cell phones. Soon, applications such as speech recognition, natural language translation, and virtual/augmented reality may leap from research prototypes to widespread commercial use.

While software has enjoyed plentiful and stable resources in the world of desktops (and to some extent, of laptops,) resource variation needs to be taken into account in smaller devices. Despite the impressive capabilities of today's mobile devices, user

J. Cordeiro et al. (Eds.): ICSOFT 2008, CCIS 47, pp. 48–61, 2009.
© Springer-Verlag Berlin Heidelberg 2009

expectations with respect to performance and sophistication will continue to be set by the full-size versions running on powerful desktops and servers.

Research in resource-adaptive applications takes an important step towards addressing resource limitation and variation [1-3].

However, existing solutions either enforce predetermined policies, or offer limited mechanisms to control the application's policies. In some cases, the adaptation mechanisms focus strictly on network conditions, enforcing policies that are established by system designers before the system is deployed. In other cases, users are offered limited control over the policies, typically focusing on a single aspect of quality of service, such as battery duration.

Unfortunately, those limitations prevent adaptive systems from addressing two important issues. First, user goals often entail *tradeoffs* among different aspects of quality of service (QoS). For example, in the presence of limited bandwidth, should a web browser skip loading pictures in order to provide faster load times? For browsing restaurant listings, a user may prefer dropping images to improve load times; but for browsing online driving directions, the user may be willing to wait longer for the full page content.

Second, user activities may involve *more than one application*, making it desirable to coordinate resource usage and adaptation policies across applications. For example, an activity that involves simultaneous video streaming and downloading email attachments may be best served when video streaming consistently uses 80% of the bandwidth and email does not attempt to go beyond 20%.

This paper presents a framework for engineering resource-adaptive systems that: (a) empower users to control tradeoffs among a rich set of aspects of QoS, and (b) coordinate resource usage among several applications. To develop such a framework, important questions need to be answered: how to represent QoS tradeoffs in a way that can be used to guide adaptation policies? How to elicit the tradeoffs preferred by users? How to allocate resources among applications, and how to coordinate their policies? What APIs must applications expose to be amenable to such coordination?

In the remainder of this paper, section 2 discusses the approach to develop such a framework, and specifically, it describes existing work on which we build. Sections 3, 4, and 5 elaborate respectively on the representation of QoS tradeoffs, on coordinating resource usage, and on enhancing applications for adaptation within the proposed framework.

Concerning evaluation, Section 3 describes a controlled user study focusing on the usability of the mechanisms for eliciting QoS tradeoffs. Sections 4 and 5 describe systems evaluations focusing on the efficiency and effectiveness of the adaptation mechanisms, respectively.

The results of the usability evaluation indicate that end-users with diverse backgrounds can understand and use the proposed model to control the adaptive behavior of applications. The results of the systems evaluations show that the implemented mechanisms are efficient, in terms of computing overhead, and effective, in the sense of making optimal decisions.

Section 6 discusses related work, and Section 7 summarizes the main points of this paper.

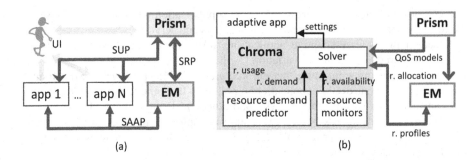

Fig. 1. The infrastructure: (a) overview of Aura, and (b) integration of Chroma

2 Architectural Baseline

Adaptability to resource limitations and variation can be promoted using two alternative architectural strategies. Either individual applications are responsible for capturing and enacting adaptation policies, or the features required to do so are factored out into a common infrastructure.

The latter approach has significant advantages both in terms of *reuse*, and of *coordinating* policies across several applications. With respect to reuse, in addition to avoiding *code* replication across applications, there is also the reuse of the *knowledge* about user preferences. For example, once the preferred QoS tradeoffs for watching a specific video stream are elicited from the user, that knowledge resides with the infrastructure and can be passed to the streaming application running on the device that happens to be convenient to the user at each moment: a cell phone, a laptop, etc.

With respect to coordination of policies, it is hard for individual applications to be aware of which other applications are being actively used, and of their resource demands and adaptation policies. Therefore, it is frequent for applications to trample each other in their quest for resources. A common infrastructure, on the other hand, may leverage a global perspective to coordinate resource allocation and policies.

Therefore, we decided to define a software infrastructure that: (a) captures models of QoS tradeoffs, (b) coordinates the resource usage across the applications supporting the user's activity, if more than one is involved, and (c) enables those applications to dynamically adjust their adaptation policies based on the models of QoS tradeoffs.

Rather than building such an infrastructure from scratch, we extended an existing infrastructure developed at Carnegie Mellon's Project Aura [4]. The remainder of this section summarizes the Aura infrastructure for ubiquitous computing [5], as well as an existing library for resource adaptation, Chroma [6], also related to Project Aura.

Aura supports a high-level notion of user activities, such as preparing presentations, or writing film reviews. Such activities may involve several services. For instance, for preparing a presentation, a user may edit slides, refer to a couple of papers on the topic, check previous related presentations, and browse the web for new developments.

Fig. 1a shows a component and connector view of the Aura infrastructure [5]. The component called Prism captures and maintains models of user activities. Specifically, each model enumerates the services required to support the activity, how those services are interconnected, if at all, preferences with the respect to the kinds of applications to provide each service (e.g., Emacs as opposed to vi for editing text,) and service-specific settings.

The Environment Manager (EM) component keeps track of the availability of services within an *environment*. An environment in Aura refers to the set of devices, software components and other resources accessible to a user at a particular location.

Whenever a user indicates that he or she wishes to start or resume an activity, Prism communicates the corresponding activity model to the EM using the service request protocol (SRP), and the two components negotiate the configuration that best supports the user's needs and preferences. Once an agreement is reached, the EM communicates with the applications using the service announcement and activation protocol (SAAP) to activate the services and make the required interconnections, if any. After that it passes a model of the concrete configuration up to Prism (SRP). Prism uses this model to communicate with the applications using the service use protocol (SUP) and recover the preferred settings for the activity; for example, the point at which the user was previously watching a video.

The Aura connectors (SAAP, SRP, and SUP), support the asynchronous exchange of XML messages over TCP/IP. These are peer-to-peer protocols, where each component may initiate communication, as needed.

Also related to project Aura, the Chroma library enables conventional applications to be enhanced for adaptation, provided applications can carry out their operations using different *tactics* [6]. For example, a speech recognizer may have sophisticated algorithms that deliver better results at the expense of high resource consumption, or simpler algorithms that demand fewer resources. Additionally, Chroma supports the partitioning of applications, shipping and running heavy computations in remote servers when the available resources, such as bandwidth, favor that option [7].

The internal structure of Chroma is outlined in the shaded area of Fig. 1b. It includes (a) monitors of available resources, (b) resource demand predictors based on application profiling, and (c) a solver for deciding which tactic to use at each moment. The thin arrows in Fig. 1b represent information flow among these components, as a result of method calls.

Currently supported resource monitors include history-based monitors of available bandwidth, battery charge, CPU and memory, both on the local device and on remote servers [8]. The resource demand predictor forecasts the resource demand of each tactic based on historical averages of actual demand.

Key to the workings of Chroma is a *utility function* encoded within the *solver* component. This utility function captures a specific resource-adaptation policy and is normally determined at design-time for each application. The solver determines the tactic with the highest utility, given the available resources, by exhaustive evaluation of all the tactics defined for the application. The solver is invoked by the application before carrying out each unit of work; for example, before recognizing each utterance, in the case of speech recognition, or before rendering each frame, in the case of virtual reality applications.

The research in this paper involved extending the Prism and EM components in Aura, as well as integrating Chroma with the Aura protocols and with the QoS models described in Section 3. The Aura protocols were also extended to include (a) the flow of QoS models from Prism to the EM, over the SRP, and to Chroma, over the SUP;

and (b) the flow of resource profiles from Chroma to the EM, and of resource allocation in the reverse direction, both over the SAAP. These flows are represented as the
thicker arrows in Fig. 1b, corresponding to the protocols in Fig. 1a.

3 Quality–of–Service Tradeoffs

Any adaptation or optimization process is guided by a goal. In the case of adapting to
resources in small mobile devices, the goal is to optimize the QoS perceived by the
user. Work in this area frequently addresses conserving resources, such as battery
charge, but that is just one way to optimize for service *duration*, an aspect of QoS.

The conceptual framework that we adopt takes into account that:

(1) Users may care about tradeoffs between different aspects of QoS; e.g., latency vs.
    accuracy.
(2) Different services may be characterized by different aspects of QoS. For example,
    for web browsing, users may care about load times and whether the full content is
    loaded (e.g., pictures); for automatic translation, users may care about the re-
    sponse time and accuracy of translation; for watching a movie, users may care
    about the frame rate and image quality.
(3) User preferences for the same service may depend on the user's activity. For ex-
    ample, a user may prefer high frame rate over image quality for watching a sports
    event over a network connection with limited bandwidth, but might prefer the op-
    posite for watching a show on sculpture.

A simple approach to modeling user preferences is to indicate which aspect of QoS a
user values the most. For example, for automatic translation, the user might indicate
that response time is preferred over accuracy, and the system could then adopt a policy that optimizes response time.

However, important questions cannot be answered with this approach: for instance,
how short of a response time will satiate the user? And even if accuracy is less important, what if it degrades so much that the translations become unusable?

At the other end of the spectrum, preferences may be expressed as an arbitrary
function between the multivariate quality space and a *utility* space representing user
happiness. For instance, the user might indicate that he would be happy with medium
translation accuracy, as long as latency remains under 1 second, and that he will be
happy to wait 5 seconds for highly accurate translations. Although fully expressive,
designing mechanisms to elicit this form of preferences from end-users is a hard problem, and even more so if more than two aspects of QoS are involved.

The model we propose sits between these two extremes. User preferences are expressed as *independent* utility functions for each aspect, or dimension, of QoS. Such
functions map the possible quality levels in the dimension to a normalized utility
space $U \equiv [0,1]$, where the user is happy with utility values close to 1, and unhappy
with utility values close to zero.

For each continuous QoS dimension the user indicates two values: the thresholds of
satiation and of starvation. For example, the user might be happy with response times
anywhere under 3 second, but may not accept response times over 20 seconds. This is

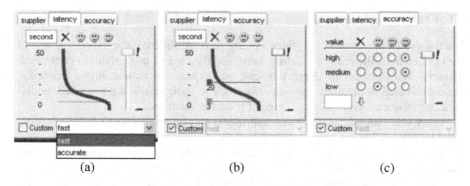

Fig. 2. QoS preferences for a language *translation* service

illustrated in Fig. 2b, where the thresholds of satiation and starvation are represented by the green (lighter) and red (darker) lines, respectively. Currently, we use sigmoid functions to smoothly interpolate between these two zones, the thresholds marking the knees of the sigmoid. The utility corresponding to each value of latency is indicated by the scale at the top, ranging from a happy face (☺) for values beyond the satiation threshold, all the way down to a cross (✗), representing rejection, for values beyond the starvation threshold.

Preferences for discrete QoS dimensions are represented using a discrete mapping to the utility space. Fig. 2c shows an example where a table indicates the utility of each level of accuracy.

The functions for each aspect of QoS are then combined by multiplication, which corresponds to an *and* semantics: a user is happy with the overall result only if he is happy with the quality along each and every dimension. Whenever a user task involves more than one service, the overall utility combines the QoS dimensions for all the services.

The relative importance of each aspect, modeled as a weight $w \in [0,1]$, is factored into the combined utility. For example, for two aspects a and b, the combined utility function is $u_a^{w_a} . u_b^{w_b}$. These weights take the value 1 by default, but may be altered using the slider bars on the right side in Fig. 2b-c.

To make it easier to use this model, we include the notion of preference *templates*. This decision is based on the principle of offering incremental benefit for incremental effort, also known as *gentle slope* systems [9]. Fig. 2a shows an example with two templates, *fast* and *accurate*. If a template is selected, the associated preferences are shown. In case a user wishes to fine-tune these preferences, he may do so after selecting the *custom* checkbox (Fig. 2b-c).

For the work herein, Prism was extended with capabilities for capturing and disseminating QoS models as illustrated above. These models are represented internally and disseminated to other components using XML, and specifically the format illustrated in Fig. 5a. The use of XML as opposed to language-specific data structures makes the models easier to exchange between components written in different languages. Prism creates user interfaces like the one in Fig. 2 dynamically, based on the XML representation of a model.

3.1 Evaluation of Usability

For the evaluation of usability, three criteria were considered: the expressiveness of the QoS models, the ease of eliciting them, and the ease of using them to control adaptation. With respect to expressiveness, our experience with multiple examples, some illustrated in the user study described below, indicate that the proposed models are expressive enough for a wide range of practical situations.

A user study investigated whether end-users can express their preferences and control adaptation using the proposed QoS models.

This study consists of using a natural language translator running on a mobile device. The quality of translation observed by users varies, since the translator runs either simple algorithms locally, or more sophisticated ones on a remote server, depending on the availability of bandwidth and of capacity in the server. To prevent limitations in the capabilities of the actual translation application (limited dictionaries, etc.) from affecting the results of the study, we replaced a human for the remote translation server. This technique is well accepted and known as a *Wizard of Oz* experiment.

The study focused on answering the following questions: first, can users understand and use templates to achieve a goal? Second, can users think of and manipulate preferences in terms of thresholds? Third, do they find it easy? And fourth, can users interpret the effects of specifying different preferences in the application's adaptive behavior?

The participants were drawn from a population with homogeneous education level and age group, but diverse technical background. Ten students in the age group 18-29 were drawn among the respondents to a posting, 5 of which from computing-related fields (computer science, electrical and computer engineering, logic) and the remaining 5 from other fields (business, physics, literature). Incidentally, 6 were male and 4 female.

Participants individually performed an experiment that lasted 30 minutes, after being given a 30 minute introduction to the experiment, methodology and tools. Participants were asked to follow the *think aloud* protocol [10], and their voice and actions on the screen were recorded using video capturing software [11]. After the experiment, the participants completed a short questionnaire.

The scenario for the experiment revolved around a conversation with a foreign language speaker (Spanish in this case) aided by translation software. To prevent serious misunderstandings in a real situation, users of the translation software would be able to check the accuracy of translation by having the Spanish translation translated back to English and spoken (using speech synthesis) on the user's earphones. Users would press a go-ahead button to synthesize the Spanish translation only if they were happy with the accuracy of translation.

During the experiment, participants were asked to input sentences of their own making, listen to the output of the double translation, and rate the accuracy. The training included calibrating the participants' rating of accuracy using the following scale: high, if the meaning is fully preserved; medium, if the meaning is roughly preserved; and low, if the meaning is seriously distorted.

Participants were asked to pursue different QoS goals during each part of a three-part experiment. Within each part, we simulated resource variation and asked the participants to evaluate the changes both in latency and accuracy of translation. During the

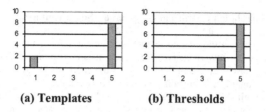

(a) Templates   (b) Thresholds

Fig. 3. Likert scale evaluation of pref. specification (5-fully favorable, to 1-unfavorable)

first two parts, the QoS goals could be satisfied by preference templates. During the third part, the specific goal could only be achieved by customized preferences. The participants were not directed as to whether or not to use templates in any case.

Whenever the QoS goals could be met by a template, the participants did use templates in 17 out of 20 cases. In the remaining 3 instances, the participants were still able to achieve the goals using customized preferences. When asked about the clarity and usefulness of templates, 8 participants were fully favorable, while 2 didn't recognize a benefit in having templates – see Fig. 3a.

All 10 participants were able to manipulate the thresholds in customized preferences for achieving the required QoS goals. Specifically, the experiment was set in such a way that the thresholds in one dimension needed to be made stricter, while relaxing the other dimension, under penalty of the goal not being achievable.

When asked about the clarity of using thresholds to specify preferences, 8 participants were fully favorable, while 2 thought some alternative strategy could be preferable – see Fig. 3b. One of these participants suggested that an X-Y representation the tradeoff might be clearer. However, there are two reasons why that may not be such a good idea. First, it would be hard to show and manipulate tradeoffs with more than two aspects of QoS. Second, the actual tradeoff changes with the availability of resources: with plentiful resources, high levels may be attainable along all aspects; but with low resources, to privilege one aspect may have a severe impact on others.

The participants were able to interpret the effects of different preferences in the application's adaptive behavior. To verify this, we tested the hypothesis that when resources change participants perceive a change in the QoS, with a greater impact along the QoS dimension for which the preferences are laxer. For that, after each translation the participants evaluated which QoS dimension changed the most relative to the previous translation: a noticeable change in accuracy with similar latencies, a noticeable change in latency with similar accuracies, no noticeable changes, etc. Participants then related those changes with the strictness or laxness of the preferences along each QoS dimension. The participants were not informed of when or in which direction resources would change.

Fig. 4 shows the results of correlating which dimension had stricter preferences with which dimension was perceived to have changed the most. The correlation coefficient is negative, meaning that whenever user preferences were stricter along one dimension, the participants perceived a greater fluctuation on the other dimension (caused by underlying resource fluctuations). When asked about how easy it was to use the interfaces in Fig. 2 to customize preferences, 5 participants were fully favorable while the other 5 thought the interfaces could be improved.

Correlation Coefficient	t-value	Significant at 95%
-0.6	-4.27	Yes

How to interpret a correlation: the correlation coefficient denotes the slope of the line that best fits the data. A positive/negative coefficient means that an increase in the x-axis corresponds to an increase/decrease in the y-axis. If the coefficient is zero, the data cannot be approximated by a straight line (there is no correlation between the x values and the y values).
Student's t-test of significance: indicates the likelihood that the correlation in the data sample corresponds to a real correlation in the general population. A commonly accepted threshold is 95% confidence. Statistics manuals contain tables of t-statistics for each size of the data sample. The t-test consists of comparing the t-value calculated for the correlation with the lookup t-statistic. If the absolute t-value is larger than the t-statistic, then the correlation is significant with 95% certainty.
Sample: 40 data points relating two variables (38 degrees of freedom), for which the t-statistic is 2.024 for a 95% confidence.

Fig. 4. Regression performed on experiment data

This user study demonstrates that end-users can both define their preferences, and interpret the results of such definitions in the system's adaptive behavior. A control loop is therefore formed, enabling users to pursue concrete QoS goals. The practicality of the control loop is confirmed by the fact that all participants were easily able to achieve concrete QoS goals.

4 Coordinating Resource Usage

For this research, Aura's EM was extended with capabilities for determining and disseminating the optimal resource allocations among the applications supporting the user's activity. For that, the EM takes three kinds of inputs: models of QoS tradeoffs via the SRP, resource profiles via the SAAP, and resource estimates.

Fig. 5a illustrates the models of QoS tradeoffs corresponding to the user preferences in Fig. 2b-c. Fig. 5b illustrates resource profiles, which relate the quality levels that each application can operate at with the corresponding resource demands. Similarly to adaptive applications (see section 5,) the EM also uses resource forecasting components. However, in contrast to the fine-grained forecasts for adaptive applications, these estimates take into account large numbers of historical samples in order to forecast availability for several seconds into the future.

Given the three inputs above, the EM computes a resource allocation for each selected application, which is optimal in the sense that it maximizes the overall utility for the user's activity. Fig. 5c shows an example of resource constraints that the EM might send to one application via the SAAP.

The EM runs the algorithm for optimal resource allocation in response to changes made by the user to the QoS models, and periodically, to address trends in resource availability.

```
                    <utility combine="product">
                      <QoSdimension name="latency" type="int">
                        <function type="sigmoid" weight="1">
                          <thresholds good="3" bad="20" unit="second"/>
                        </function>
                      </QoSdimension>
(a)                   <QoSdimension name="accuracy" type="enum">
                        <function type="table" weight="1">
                          <entry x="high" f_x="1"/>
                          <entry x="medium" f_x="1"/>
                          <entry x="low" f_x="0.3"/>
                        </function>
                      </QoSdimension>
                    </utility>

                    <service type="speechRecognition">
                      <QoSprofile>
                        <QoSdim name="latency" type="float"/>
                        <QoSdim name="accuracy" type="enum"/>
                        <head>latency accuracy cpu bdwdth</head>
(b)                     <units>second none    %    Kbps</units>
                        <point> 0.05  low      30     250</point>
                        <point> 0.05  high     80     250</point>
                        <point> 0.1   low      20     200</point>
                        <point> 0.1   high     75     200</point>
                      </QoSprofile>
                    </service>

                    <constraints>
(c)                   <rsrc id="cpu" avg="30" var="10" u="%"/>
                      <rsrc id="bdwdth" avg="800" var="100" u="Kbps"/>
                    </constraints>
```

Fig. 5. (a) Representation of the preferences in Fig. 2; (b) example QoS profile; (c) example resource allocation

4.1 Evaluation

EM's evaluation focused on efficiency, making sure that running the resource alloca-tion algorithm does not introduce perceptible delays for the user, and does not draw significantly from the resources available to the applications. The experiments were carried out on an IBM ThinkPad 30 laptop running Windows XP Professional, with 512 MB of RAM, 1.6 GHz CPU, and WaveLAN 802.11b card. Prism and the EM each run on a Hot Spot JRE from Sun Microsystems, version 1.4.0_03.

The average latency finding the optimal configuration is 200 ms (standard devia-tion 50 ms) for user tasks requiring from 1 to 4 services, when 4 to 24 alternative suites of application are available to provide those services, and when the search space of combined QoS levels reaches up to 15,000 points.

The memory footprint of the EM ranges linearly from 7 MB to 15 MB when it holds the descriptions of 20 up to 400 services in the environment. By comparison, a "hello world" Java application under the used Java Runtime Environment (JRE) has a memory footprint of 4.5 MB, and a Java/Swing application that shows a "hello world" dialog box has a memory footprint of 12 MB.

When reevaluating the resource allocation every 5s, the EM uses on average 3% of CPU cycles. The optimality of decisions was verified analytically, and more details about the EM's evaluation can be found in [12].

5 Adaptive Applications

The key requirements for adaptive applications within the proposed infrastructure are: first, to comply with the resource allocation determined by the EM; and second, to enforce the QoS tradeoffs communicated by Prism, in the face of resource variations.

To reduce the costs of addressing these requirements in every application, we decided to integrate Chroma, a software layer for resource adaptation [6]. Architecturally, adaptive applications are built on top of Chroma, and there is one run-time instance of the Chroma library deployed with each application.

Integrating such applications involved wrapping them to mediate between the protocols supported by the Aura connectors and the Chroma APIs. The thicker arrows in Fig. 1b represent information flow between Prism, the EM, and Chroma, over the Aura connectors. Since Chroma expects a generic utility function to be encoded within the solver, plugging in a function that interprets the QoS models passed via the SUP (Fig. 5a) was fairly straightforward.

In the interest of space, how to enhance applications for adaptation with Chroma is not further discussed here, but details can be obtained in [6].

5.1 Evaluation

The systems evaluation presented here follows closely the experiment described in Section 3.1. This test is based on the scenario where a PDA is used to carry out natural language translation. When resources are *poor,* no remote servers can be reached, and the translation is carried out exclusively using the PDA's capabilities. When resources are *rich,* powerful remote servers are available to do part of the work. We used a 233Mhz Pentium laptop with 64MB of RAM to simulate the PDA and 1GHz Pentium 3 laptops with 256MB of RAM as the remote servers.

The test used 3 randomly selected sentences, of between 10-12 words in length. Each sentence was (doubly) translated five times, from English to Spanish and then back into English using Pangloss-Lite, a language translation application [13].

The utility functions provided to Chroma correspond to the *fast* and *accurate* templates introduced in Section 3. The fast template accepts medium accuracy within 1s, and the *accurate* template is willing to wait 5s for highly accurate translations. The latency thresholds for the system testing are much smaller than the ones used in the user study in Section 3.1, in which the translation was performed by a team member. Each sentence was translated under rich and poor resources, for each of the two preference templates (four test situations).

Table 1 shows the relative utility of Chroma's decisions in each of the four test situations. The relative utility is calculated as the utility of the QoS delivered by Chroma's decision relative to the best possible utility, among all the alternative tactics, given the

Table 1. Relative utility of Chroma's decisions

preferences / resources	Chroma's decision	Lower resources	Higher resources
fast / poor	1.0	N/A	0.45
accurate / poor	1.0	0.37	0.13
fast / rich	1.0	0.51	0.83
accurate / rich	1.0	0.50	N/A

current resource conditions. To illustrate this optimality, the two rightmost columns show the utility of the adjacent decisions in terms of resource usage. That is, the relative utility of the decisions that would take the nearest lower resources, and the nearest higher resources, respectively. There are two corner cases, shown with N/A, where Chroma's decision corresponds to the lowest possible resource usage, and to the highest possible usage.

In summary, Chroma always picks the best possible tactic, under different resource conditions, and different user preferences. A more thorough validation of Chroma's ability to perform adaptation in the presence of limited resources is presented in [7].

6 Related Work

Similarly to the proposed framework, others have leveraged techniques from micro-economics to elicit utility with respect to multiple attributes. In the Security Attribute Evaluation Method (SAEM), the aggregate threat index and the losses from successful attacks are computed using utility functions [14]. The Cost Benefit Analysis Method (CBAM) uses a multidimensional utility function with respect to QoS for evaluating software architecture alternatives [15]. Our work is different from SAEM and CBAM in that it is geared towards mobile computing.

A body of work addressed battery duration in mobile devices. For example [1], presented OS extensions that coordinate CPU operation, OS scheduling, and media rendering, to optimize device performance, given user preferences concerning battery duration. The QoS models in our framework are significantly more expressive, since they support a rich vocabulary of service-specific aspects of QoS.

User studies done in mid-to-late 1990s have demonstrated that stability (e.g., absence of jitter) is more important than improvement for certain aspects of QoS [16]. Our framework recognizes the importance of these results and ensures, by explicit resource allocation, that adequate resources are available for applications to provide service while maximizing the overall utility.

Dynamic resolution of resource allocation policy conflicts involving multiple mobile users is addressed in [17] using sealed bid auctions. While this work shares utility-theoretic concepts with our configuration mechanisms, the problem we solve is different. Our work has no game-theoretic aspects and addresses resource contention by multiple applications working for the same user on a small mobile device.

From an analytical point of view, closest to our resource allocation algorithm are Q-RAM [18], Knapsack algorithms, and winner determination in combinatorial auctions. By integrating with generic service discovery mechanisms in the EM, our work provides an integrated framework for service discovery, resource allocation and adaptation [5].

7 Conclusions

Resource adaptation can play an important role in improving user satisfaction with respect to running sophisticated software on small mobile devices.

However, today, many applications implement limited solutions for resource adaptation, or none at all. The primary reasons for that are: (a) the cost of creating ad-hoc adaptation solutions from scratch for each application; and (b) the difficulty of coordinating resource usage among the applications. Because it is hard for an individual application to even know which other applications are actively involved in supporting a user's activity, individual applications frequently trample each other in their quest for resources.

This paper proposes a framework for resource adaptation where a number of features are factored out of applications into a common infrastructure.

First, user preferences with respect to overall QoS tradeoffs are elicited by an infrastructural component such as Prism. These models are expressed using a rich vocabulary of service-specific QoS aspects.

Second, resource allocation among applications is coordinated by another infrastructural component such as the EM. This component receives QoS profiles from applications, and efficiently computes the resource allocations that optimally support the QoS goals, given forecasts of available resources for the next few seconds.

Third, adaptation to resource variations at a time granularity of milliseconds is facilitated by a common library, such as Chroma. This library saves application development costs by providing common mechanisms for (a) monitoring available resources, (b) profiling the resource demands of alternative computation tactics, and (c) deciding dynamically which tactic best supports the QoS goals, given resource forecasts for the next few milliseconds.

Additionally, this paper clarifies concrete APIs that adaptive applications need to support for being integrated into the framework. These APIs are realized as XML messages, which may be exchanged within the mobile device, or across the network, if some of the infrastructural components are deployed remotely.

In summary, the proposed framework makes it easier to develop and integrate applications into coordinated, resource-adaptive systems. Furthermore, our research indicates that end-users with diverse backgrounds are able to control the behavior of such systems to achieve activity-specific QoS goals.

References

1. Yuan, W., Nahrstedt, K., Adve, S., Jones, D., Kravets, R.: GRACE-1: Cross-Layer Adaptation for Multimedia Quality and Battery Energy. IEEE Transactions on Mobile Computing 5, 799–815 (2006)
2. De Lara, E., Wallach, D., Zwaenepoel, W.: Puppeteer: Component-based Adaptation for Mobile Computing. In: USENIX Symposium on Internet Technologies and Systems (USITS), pp. 159–170. USENIX Association, San Francisco (2001)

3. Flinn, J., Satyanarayanan, M.: Energy-aware Adaptation for Mobile Applications. ACM SIGOPS Operating Systems Review 33, 48–63 (1999)
4. Garlan, D., Siewiorek, D., Smailagic, A., Steenkiste, P.: Project Aura: Toward Distraction-Free Pervasive Computing. IEEE Pervasive Computing 1, 22–31 (2002)
5. Sousa, J.P.: Scaling Task Management in Space and Time: Reducing User Overhead in Ubiquitous-Computing Environments. Carnegie Mellon University, Pittsburgh (2005)
6. Balan, R.K., Gergle, D., Satyanarayanan, M., Herbsleb, J.: Simplifying Cyber Foraging for Mobile Devices. Carnegie Mellon University, Pittsburgh (2005)
7. Balan, R.K., Satyanarayanan, M., Park, S., Okoshi, T.: Tactics-Based Remote Execution for Mobile Computing. In: USENIX Intl. Conference on Mobile Systems, Applications, and Services (MobiSys), pp. 273–286. ACM, San Francisco (2003)
8. Narayanan, D., Flinn, J., Satyanarayanan, M.: Using History to Improve Mobile Application Adaptation. In: 3rd IEEE Workshop on Mobile Computing Systems and Applications (WMCSA), Monterey, CA (2000)
9. Myers, B., Smith, D., Horn, B.: Report of the End-User Programming Working Group. In: Myers, B. (ed.) Languages for Developing User Interfaces, pp. 343–366. Jones and Barlett, Boston (1992)
10. Steinberg, E. (ed.): Plain language: Principles and Practice. Wayne State University Press, Detroit, MI (1991)
11. TechSmith: Camtasia Studio (accessed 2008), http://www.techsmith.com/
12. Poladian, V., Sousa, J.P., Garlan, D., Shaw, M.: Dynamic Configuration of Resource-Aware Services. In: 26th International Conference on Software Engineering, pp. 604–613. IEEE Computer Society, Edinburgh (2004)
13. Frederking, R., Brown, R.: The Pangloss-Lite Machine Translation System. In: Expanding MT Horizons: Procs 2nd Conf Association for Machine Translation in the Americas, Montreal, Canada, pp. 268–272 (1996)
14. Butler, S.: Security Attribute Evaluation Method. A Cost-Benefit Approach. In: Intl. Conf. in Software Engineering (ICSE), pp. 232–240. ACM, Orlando (2002)
15. Moore, M., Kazman, R., Klein, M., Asundi, J.: Quantifying the Value of Architecture Design Decisions: Lessons from the Field. In: Intl. Conf. on Software Engineering (ICSE), pp. 557–562. IEEE Computer Society, Portland (2003)
16. Wijesekera, D., Varadarajan, S., Parikh, S., Srivastava, J., Nerode, A.: Performance evaluation of media losses in the Continuous MediaToolkit. In: Intl. Workshop on Multimedia Software Engineering (MSE), Kyoto, Japan, pp. 60–67. IEEE, Los Alamitos (1998)
17. Capra, L., Emmerich, W., Mascolo, C.: CARISMA: Context-Aware Reflective mIddleware System for Mobile Applications. IEEE Transactions on Software Engineering 29, 929–945 (2003)
18. Lee, C., Lehoczky, J., Siewiorek, D., Rajkumar, R., Hansen, J.: A Scalable Solution to the Multi-Resource QoS Problem. In: IEEE Real-Time Systems Symposium (RTSS), pp. 315–326. IEEE Computer Society, Los Alamitos (1999)

On the Multiplicity Semantics of the Extend Relationship in Use Case Models

Miguel A. Laguna and José M. Marqués

Department of Computer Science, University of Valladolid
Campus M. Delibes, 47011 Valladolid, Spain
{mlaguna,jmmc}@infor.uva.es

Abstract. Use cases are a useful and simple technique to express the expected behavior of an information system in successful scenarios or in exceptional circumstances. The weakness of use cases has been always the vague semantics of the relationships, in particular the *extend* relationship. The main contribution of this article is an attempt to clarify the different interpretations that can be adopted. A major revision of the UML standard would be impractical, but the *extension point* concept could be completed, including minimum and maximum multiplicity attributes. Using these minor changes, the legal combination of base/extending use cases in the requirements models would be unequivocally defined. Therefore, the ambiguity of the original UML models would be removed.

1 Introduction

Use cases are one of the preferred techniques for the elicitation and definition of the intended behavior of the system under study. They are a useful and simple technique to describe the successful scenarios (where things occur as expected) or the problematic situations (alternative and exceptional paths). Use cases were an original idea of Jacobson, incorporated in his OOSE development method [12]. From the first versions of UML as standard modeling language [17], use cases have been chosen as the preferred technique to identify and define the user requirements and to represent the behavior of the system as a black box, in place of other techniques used until then (for example, the Rumbaugh OMT method [19] used data flow diagrams). They are basic in the Unified Process, as this was evolved from the ideas of Jacobson [11]. Many criticisms have been made concerning use cases; see for example the articles of Berard [1], Simons [21], or more recently Isoda [9]. Conversely, there are many works that try to improve or at least clarify them, such as the classical book of Cockburn [3] or the work of Williams *et al.* [22].

Some authors have suggested that the most important characteristics of use cases are the textual details to be discussed with the end users while neglecting the visual representation and semantics proposed by UML. Others, such as Rumbaugh and Jacobson, continue to promote the graphics aspects [20]. Constantine connects user interface design methods with the use case elicitation and refinement [4]. Some additional relationships and other different meta-model modifications are proposed. More details about these questions can be found in the related work section.

J. Cordeiro et al. (Eds.): ICSOFT 2008, CCIS 47, pp. 62–75, 2009.

One of the major controversies is the UML's explanations of *include* and *extend* relationships. These concepts remain vague, and apparently contradictory, confusing readers (and also some authors of software engineering books) about when to use *include* or *extend*. Precise and unambiguous definitions of terms are missing in the numerous UML documents. Therefore, UML's explanations for *include* and *extend* relationships are still subject to ongoing debate. Some conferences have been devoted to these and other conflicting aspects [5].

The rest of the paper is as follows: The next section briefly summarizes the evolution of *include* and *extend* relationships in UML documents. Sections 3 and 4 specifically discuss the problems with the *extend* relationship and propose some semantic reinterpretations and minor meta-model modifications. Section 5 presents related work and section 6 concludes the paper and proposes additional work.

2 The Evolution of the Extend Relationship

It is well known that a use case describes an interaction between one or more actors and the system as a sequence of messages. Thus, a use case diagram has two types of nodes: actors and use cases, connected by association relationships. The original proposal of Jacobson also included two kinds of relationships between use cases: The *uses* and *extends* relationships, indicated with generalization arrows. This syntax was initially preserved in primitive UML versions (see Figure 1) but, beginning with the refined 1.3 version, a new set of relationships was proposed and this definition has essentially been kept, with minor changes, until the actual UML 2.1.2 version.

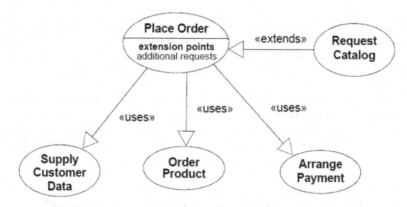

Fig. 1. The syntax of the old *extends* and *uses* relationships, as exemplified in the 1.1 version of UML [17]

From UML 1.3, relationships between use cases can be expressed in three different ways: with generalization, *include*, and *extend* relationships (see Figure 2 for *extend* and *include* examples):

- A generalization relationship between use cases implies that the child use case contains the behavior of the parent use case and may add additional behavior.

- An *include* relationship means that the behavior defined in the target use case is included at one location in the behavior of the base use case (it performs all the behavior described by the included use case and then continues with the original use case).
- An *extend* relationship defines those instances of a use case that may be augmented with some additional behavior defined in an extending use case.

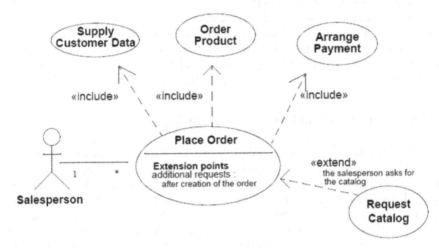

Fig. 2. The syntax of the actual *extend* and *include* relationships, as they appear in the 2.1.2 version of UML [16]

The semantics of *include* relationship has always been reasonably clear. However, the *extend* relationship has generated a lot of controversy. The variety of diverse interpretations that different authors use in textbooks or research papers is surprising, but it is less surprising if we read some fragments of the description of the UML 1.3 "clarifying" description:

"An *extend* relationship defines that a use case may be augmented with some additional behavior defined in another use case. [...] The *extend* relationship contains a condition and references a sequence of *extension points* in the target use case. [...] Once an instance of a use case is to perform some behavior referenced by an *extension point* of its use case, and the *extension point* is the first one in an *extend* relationship's sequence of references to *extension points*, the condition of the relationship is evaluated. [...] Note that the condition is only evaluated once: at the first referenced *extension point*, and if it is fulfilled all of the extending use case is inserted in the original sequence. An *extension point* may define one location or a set of locations in the behavior defined by the use case. However, if an *extend* relationship references a sequence of *extension points*, only the first one may define a set of locations. [...]"

Several modifications have been added to the different versions of UML. Attempts at removing these difficulties have been proposed in these documents. From here until the end of the article, we base the discussion on the official UML documentation, version 2.1.2 [16]. Figure 3 shows the Use Case Package of UML 2.1.2 superstructure meta-model.

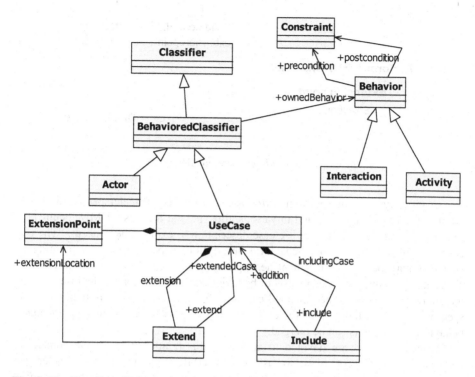

Fig. 3. The Use Case Package of UML 2.1.2 Superstructure meta-model and the associated Behavior [16]

In the UML 2.1.2 meta-model, *Actor* and *UseCase* are both *BehavioredClassifier*, which itself is a descendent of *Classifier*. This is problematic for use cases, as a use case describe a set of interactions more than a set of instances. Other authors have analyzed this question, in particular if a use case is a classifier with an associated behavior or in fact must be a behavior itself [6]. Some changes have been incorporated from version 2.0 to 2.1. *Actor* in UML 2.0 was simply a *Classifier*, not a *BehavioredClassifier*. These variations make it difficult to understand the semantics of the meta-model.

As UML documentation states, the *extend* relationship specifies how and when the behavior defined in the extending use case can be inserted into the behavior defined in the extended use case (at one *extension point*). Two important aspects are: a) this relationship is intended to be used when some additional behavior can be added to the behavior defined in another use case; b) the extended use case *must be independent* of the extending use case.

Analyzing the meta-model, the *extensionLocation* association end references the *extension points* of the extended use case where the fragments of the extending use case are to be inserted. An *extensionPoint* is an owned feature of a use case that identifies a point in the behavior of a use case where it can be extended by another use case. The *extend condition* is an optional *Constraint* that references the condition that must hold for the extension to take place. The notation for conditions has been changed in UML 2: the condition and the referenced *extension points* is included in a Note attached to the *extend* relationship (Figure 4).

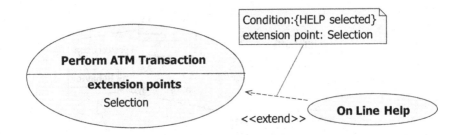

Fig. 4. The *extend* and *condition* representation in UML 2.1.2 [16]

Semantically, the concept of an *"extension location"* is left underspecified in UML because use cases "are specified in various idiosyncratic formats". UML documentation refers to the typical textual use case description to explain the concept: "The use case text allows the original behavioral description to be extended by merging in supplementary behavioral fragment descriptions at the appropriate insertion points". Thus, an extending use case consists of behavior fragments that are to be inserted into the appropriate spots of the extended use case. An extension location, therefore, is a specification of all the various (extension) points in a use case where supplementary behavioral increments can be merged.

The next sections are devoted to analyzing this relationship and the connected *extension point* concept. First, we assume the UML meta-model and consider the different semantic interpretations of the extension concept and the way the ambiguity can be removed. Then, in section 4, we discuss the necessity of the *extension point* concept itself.

3   The Interpretation of the Extension Point Concept

In this section, we assume that the extension point is a valuable concept and analyze the different possible interpretations, trying to remove ambiguity. Consider the typical example of Figure 5, where a Process Sale use case has an *extension point* Payment and several use cases *extend* the use case at this point.

The question is: What exactly does the *extension point* Payment mean? Is it a blank space that must be compulsorily refilled? And, if this is true, is it correct to add the behavior of only one of the three use cases or is it legal to add the consecutive behavior of two of these? For instance, if one and only one of the use cases must be selected, we really have a sort of polymorphism, as Figure 6 tries to show. Really, the syntax of the figure is correct from the point of view of UML 2. The imagination of the modeler can add the rest: the fragments of the Cash/Credit/Check Payments use cases can substitute the sort of interface that the Process Payment use case represents, and this last is needed to complete the behavior of the Process Sale use case.

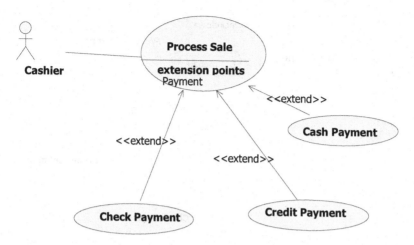

Fig. 5. The Use Case Process Sale, extended by three alternative use cases

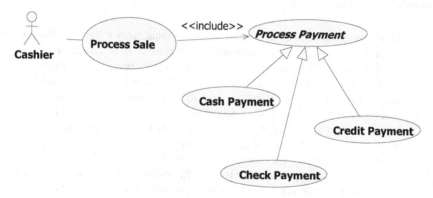

Fig. 6. The Use Case Process Sale and a possible interpretation of the Process Payment extension

We think that it is necessary to clarify the different possibilities that can appear in a system:

1) The situation is well established from the very beginning, as in the preceding example. The requirements for a simple store could be "All sales imply a method of payment" and "Only one payment method can be authorized". In these over-simplified situations, Figure 6 states clearly the semantics of the real behavior better than the pure *extend* relationship.

2) The situation is well established, extension is mandatory but flexible. The requirements could be: "All sales imply at least one method of payment". The problem now is that we cannot directly express this difference in the diagram. An illegal (i.e., not present in the UML meta-model) multiplicity annotation in the include relationship could help (see the interpretation of Figure 7). Otherwise, a change of the *include* relationship from a stereotyped dependence to an association could solve the problem. Really, the evolution of the original *uses* relationship to a dependence relationship with the new name *include* was a conflicting choice in the old UML1.1 to UML1.3 transition time.

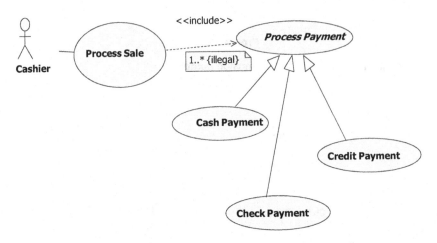

Fig. 7. The Use Case Process Sale and a reinterpretation of the Process Payment extension as a relationship with explicit multiplicity

3) Other situations can be predictable but not mandatory ("the SalesPerson can ask for the catalog" as in the example of Figure 2). In this case, the semantics correspond to an optional behavior in a specific point of the *extend*ed use case ("after creation of the order"). We need here all the assortment of details: *extension point* declaration, extending use case and *constraint*.

4) The last possibility, nearest to the original *extend* semantics, is that the situation that we want to solve is completely open in an unexpected way. In this case, the mere inclusion of an *extension point* in the "perhaps may be extended in the future" use case is contradictory. We do not know if any step of the use case description will have an alternative path a posteriori. The proposal is that we do not need any *extension point*; the "may be extended" use case must be able to be added as a special step in the exception/alternative paths set. This links to the next section's discussion and the solution proposed there: do not specify *extension points* (we cannot do it in any case as we cannot anticipate all the possible behavior modifications). The interpretation can be made explicit with the example in Figure 8. A new use case is added, based (via a generalization relationship) on the original unchanged use case. This new version has all the steps of the old use case and the new *extension point*. Now, the additional behavior can be connected via the *extend* relationship. As in the first and third variants, only the possibilities of the current UML meta-model are exploited. However, the problematic of the three possible variants considered in the previous situations (always one extension, at least one, zero or more) must be solved.

Summing up, we can use the elements of the UML meta-model to specify most of the situations, except for that stated in the second point (mandatory but flexible extension). We reach a crossroads. The radical proposal would be to modify completely the UML use case package, clarifying its general semantics and syntax (and this is a long awaited demand of many requirements specialists, as the related work section will make clear). The pragmatic possibility is to keep the actual Use Case Package, while suggesting minor changes. This implies facing two different problems: well known extensible situations (this problem refers to situations 1, 2 and 3) and unpredictable extensions (situation 4). Solving the first problem, the second is solved in two steps, as explained above, following the scheme of Figure 8.

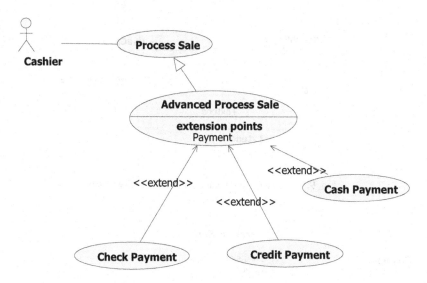

Fig. 8. The Use Case Process Sale and a reinterpretation of the Process Payment extension as an a posteriori addition

To solve the first set of situations, removing any ambiguity from the visual representation of the model, we need to complete the diagram with multiplicity details: The proposal consists of minimally modifying the UML meta-model, adding a generalization relationship from *ExtensionPoint* to *MultiplicityElement* from the Multiplicity Package. This solution implies that the *ExtensionPoint* meta-class would now have the lower and upper attributes (Figure 9).

The advantage of this solution is that the meta-model is not essentially changed. But the *extension point* would have additional and clarifying information, which allows us to assign an integer value to the new *lower* and *upper ExtensionPoint* attributes:

- 0..1 multiplicity states that the extending use case is added only in certain circumstances (when the constraint condition is true). This is equivalent to the actual semantic interpretation given by UML documentation.
- 1..1 multiplicity states that one of the possibly n *extend* use cases can be inserted. At the same time, the constraint conditions of each *extend* use case cannot overlap (See Figure 10).
- 1..n (with n>1) or 1..* multiplicity allows more than one use case to add behavior to the original use case (in our example, two consecutive payment kinds can be authorized).

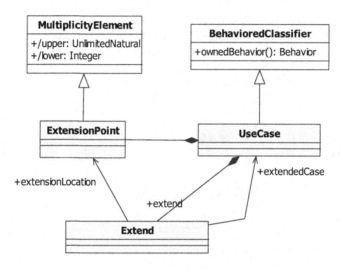

Fig. 9. The Use Case Package with multiplicity added

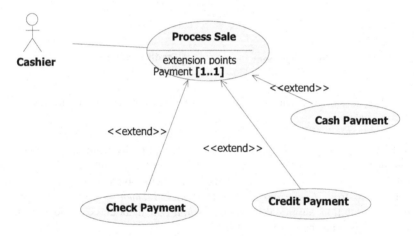

Fig. 10. The Use Case Process Sale, extended by three alternative use cases

The remaining possibility (situation 4, open to extension in any unexpected way) can be handled using the generalization relationship, as in Figure 8, combined with the modified semantics of the *extension point*. We believe that the combination of the two interpretations covers all the practical situations and solves the problems that the requirements engineers face in their daily work.

4 Discussion

The previous section has shown that the use of the *extension point* concept is problematic and must be dealt with carefully. In this section, we try to answer an earlier question: Is the presence of the *extension point* concept in the use case models really indispensable? From the point of view of the semantics of the dependence relationship, the mere presence of an *extension point* in the base use case is confusing. To remove (or perhaps to reinterpret) the *extension point* concept could perhaps be a way of avoiding many problems.

The first intention of a dependence relationship is to establish a directed relationship between an independent element (the base or extended use case) and a dependent element (the extending use case). Therefore, if the base use case must have no information a priori about the extending use case, the obligation of predetermining an *extension point* is contradictory. The well known open-closed principle states that (generally speaking) a piece of software must be completely closed from the point of view of the existing *clients* (in this case, the rest of software artifacts: classes, sequence diagrams or simply requirements documentation artifacts) and open to possible enhancements for new *clients* (new requirements or enhancements). This idea typically applies to inheritance relationships between classes in object oriented designs but can also be adopted in requirements artifacts.

The types of problems we want to solve are, for example: a use case can evolve during the development of several versions of a software system; the requirements can change; new constraints or business rules can appear, etc. The essence of these situations is that the evolution usually occurs "in an unexpected way". While the user requirements are being elicited, we have a possible solution with plain use cases: add an alternative sequence of steps to the set of exceptions of the use case, referring to a step of the main scenario. The generalization of the idea is exactly the extension concept, useful when a) the use case is already completely developed through a collaboration that involves analysis or design models, or b) the complexity of the steps that must be added recommends separating this piece of behavior in a new use case. In both cases, as in the plain solution, we must be able to indicate where the new sequence must be inserted (after the original step n) and where the original scenario must continue (after the original step m). This can be as complex as needed, as in the idea of *extension points* with several fragment insertions.

Surprisingly, the concept of step is not directly present in the UML meta-model Use Case Package, probably in order to allow different particular implementations (visual or textual, formal, structured or informal). Really, a *BehavioredClassifier* has an associated *Behavior* that can have a set of atomic *actions* or *states* ... and this could be identified as the steps of the sequence of messages of the original textual use cases. However, independently of the concrete format, the concept of sequence of steps should have to be present (or specialized as in other packages) in this meta-model Package, as Figure 11 suggests.

As we do not foresee immediate changes in the UML meta-model, we propose an apparently incorrect solution to deal with this problem: consider that a use case has a set of steps (or sequence of inseparable steps) called *extension points*. In fact, observing Figure 11, new *Step* and *ExtensionPoint* have similar relationships with *UseCase* and *Extend* meta-classes. If we think this way, quite simply, all the steps of a use case are extensible. This interpretation implies that the use cases are completely open to future extensions (in the same way an unaffected class can be extended by a new one

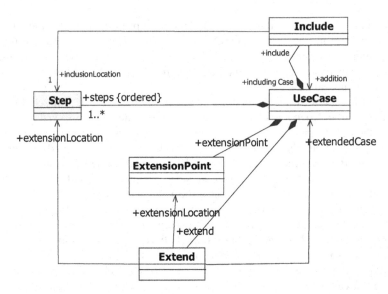

Fig. 11. The Use Case step concept in an UML hypothetical extension

using inheritance in object oriented languages). Really, our intention is only conceptual: the details are in the textual step-based description of the use cases. In practical terms, this supposes that the *extension point* concept is not used in the diagrams. In the textual documentation of the extending use case, we must indicate:

a) The use case modified.

b) The fragment/step where the extended use case is modified, using the same conventions of the alternative/exception fragments of the monolithic use cases; in other words, the precise step number must be referred.

c) The "return point" of the extended use case in order to continue with the normal sequence of steps.

The adoption of this approach means that all the possible situations must be documented in the textual information of the extending use case. The extended use case remains unchanged and unaware of the extensions.

Summarizing the idea, in many cases (in particular in agile developments), it is preferable not to use *extension points* with the original UML semantics (or the modified version suggested in this article). Or, changing the point of view, all the steps of a use case can be considered as *extension points*. This version smooths the learning curve of the technique by beginners (in fact we use this approach with our undergraduate students, avoiding many confusing discussions in the requirements gathering sessions).

5   Related Work

Many criticisms of and suggestions for modification of the UML meta-model have been proposed, including the use of ontologies instead [7]. Some additional relationships between use cases have been proposed, such as the *precedes* relationships from the

OPEN/OML method [8]. Rosenberg [18] uses the *precedes* and also the *invokes* constructs to factor out common behavior. Conversely, other authors such as Larman [14] advocate not using the *extend* relationship or using only when it is undesirable to modify the base use case.

The *BehavioredClassifier* specialization of the use cases has been analyzed in [6]: The *Behavior* meta-class is a specification of how its context classifier (use case) changes over time and the *BehavioredClassifier* is a classifier that can have behavior specifications. In other words, a *BehavioredClassifier* is rather an ordinary classifier that can own behaviors [6]. The conclusion is that the formalization of use cases as classifiers in UML has obscure points: Two contradictory notions of use cases coexist in UML 2: "set of interactions" vs. "set of entities". The authors propose the meta-model should be changed to make *UseCase* a subtype of *Behavior*, not of *BehavioredClassifier*. Alternatively, they admit that the meta-model may be kept as it is, but it should be recognized that a use case is the specification of a role. Williams *et al.* also analyze the UML 2 meta-model and propose changing *UseCase* to a subclass of *Behavior* [22].

Isoda states that UML 2 has a correction about the relationship between use cases and actors, which effectively means that UML has finally abandoned the idea of "actors call operations of a use case", but the details of UML 2 in fact still retain those defects [9].

Jacobson believes that integrating use cases and aspect oriented programming (AOP) will improve the way software is developed. The idea is to slice the system and keep the use cases separate all the way down to the code. "In the long term we will get more of extension-based software-extensions from requirements all the way down to code and runtime; and extensions in all software layers, for example, application, middleware, systemware, and extensions across all these layers" [10].

Braganza *et al.*, discuss the semantics of use case relationships and their formalization using activity diagrams in the context of variability specification. They propose an extension to the *extend* relationship that supports the adoption of UML 2 use case diagrams into model driven methods. The proposal results from the 4 Step Rule Set, a model driven method in which use cases are the central model for requirements specification and model transformation [2].

The common conclusion of most of the work done in use case semantics is that the question is not well solved in UML and a redefinition of the concepts is needed. We believe that our contribution can help in this redefinition.

6   Conclusions and Future Work

In this article, the problems of interpretation of the *extend* semantics in use case models are analyzed. The possible situations are studied and an interpretation is given for each of them. A possible improvement of the *extension point* concept is proposed, assuming that the use of this construction is useful in certain circumstances. The multiplicity attributes added to the extension point suppose a clarification of the expected behavior it is possible to add in those places. We think that, without neglecting major future modifications in the UML meta-model, this slight change can help in the process of elicitation and specification of functional requirements, clarifying the intention of the final users.

We have implemented the modified meta-model (really the Ecore version of UML meta-model) with the GMF/Eclipse platform. The building of a set of experimental mini-CASE tools (we are only interested in the use case diagrams) is a work in process to check the usefulness of the approach. The intention is to use this tool with undergraduate students and validate the comprehension of the multiplicity attribute in the *extension point* concept.

Acknowledgements. This work has been supported by the Junta de Castilla y León project VA-018A07.

References

1. Berard, E.V.: Be Careful with Use Cases. Technical report (1995)
2. Braganca, A., Machado, R.J.: Extending UML 2.0 Metamodel for Complementary Usages of the «extend» Relationship within Use Case Variability Specification. In: Proceedings of the 10th international on Software Product Line Conference, pp. 123–130. IEEE Computer Society, Washington (2006)
3. Cockburn, A.: Goals and Use Cases. Journal of Object Oriented Programming, 35–40 (September 1997)
4. Constantine, L., Lockwood, L.: Software for Use. Addison-Wesley, Reading (1999)
5. Génova, G., Llorens, J., Metz, P., Prieto-Díaz, R., Astudillo, H.: Open Issues in Industrial Use Case Modeling. In: Jardim Nunes, N., Selic, B., Rodrigues da Silva, A., Toval Alvarez, A. (eds.) UML Satellite Activities 2004. LNCS, vol. 3297, pp. 52–61. Springer, Heidelberg (2005)
6. Génova, G., Llorens, J.: The Emperor's New Use Case. Journal of Object Technology 4(6), 81–94 (2005); Special Issue: Use Case Modeling at UML-2004
7. Genilloud, G., William, F.: Use Case Concepts from an RM-ODP Perspective. Journal of Object Technology 4(6), 95–107 (2005); Special Issue: Use Case Modeling at UML-2004
8. Henderson-Sellers, B., Graham, I.: The OPEN Modeling Language (OML) Reference Manual. SIGS Books, New York (1997)
9. Isoda, S.: A Critique of UML's Definition of the Use-Case Class. In: Stevens, P., Whittle, J., Booch, G. (eds.) UML 2003. LNCS, vol. 2863, pp. 280–294. Springer, Heidelberg (2003)
10. Jacobson, I.: Use Cases and Aspects—Working Seamlessly Together. Journal of Object Technology (July/August 2003), http://www.jot.fm
11. Jacobson, I., Booch, G., Rumbaugh, J.: The Unified Software Development Process. Addison-Wesley, Reading (1999)
12. Jacobson, I., Christerson, M., Jonsson, P., Overgaard, G.: Object-Oriented Software Engineering, A Use Case Driven Approach. Addison Wesley, Reading (1994)
13. Jacobson, I., Griss, M., Jonsson, P.: Software Reuse. Architecture, Process and Organization for Business Success. ACM Press/ Addison Wesley/ Longman (1997)
14. Larman, C.: Applying UML and Patterns: An Introduction to Object-Oriented Analysis and Design and the Unified Process, 3rd edn. Addison Wesley, Reading (2004)
15. Object Management Group (OMG), Reusable Asset Specification (RAS), ptc/04-06-06 (2004)
16. OMG, Unified Modeling Language: Superstructure, version 2.1.2. Formal doc. 2007-11-02 (2007)
17. Rational Software Corporation, Unified Modelling Language Version 1.1 (1997)

18. Rosenberg, D., Scott, K.: Applying Use Case Driven Object Modeling with UML: A Practical Approach. Addison Wesley, Reading (1999)
19. Rumbaugh, J., Blaha, M., Premerlani, W., Eddy, F., Lorensen, W.: Object-Oriented Modeling and Design. Prentice Hall, Englewood Cliffs (1991)
20. Rumbaugh, J., Jacobson, I., Booch, G.: The Unified Modeling Language Reference Manual, 2nd edn. Addison-Wesley Professional, Reading (2004)
21. Simons, A.J.H.: Use Cases Considered Harmful. In: 29th Conf. Tech. Obj.-Oriented Prog. Lang. and Sys. (TOOLS-29 Europe). IEEE Computer Society, Los Alamitos (1999)
22. Williams, C., Kaplan, M., Klinger, T., Paradkar, A.: Toward Engineered, Useful Use Cases. Journal of Object Technology 4(6), 45–57 (2005); Special Issue: Use Case Modeling at UML-2004

Secure Mobile Phone Access
to Remote Personal Computers: A Case Study

Alireza P. Sabzevar and João Pedro Sousa

Computer Science Department, George Mason University
4400 University Drive 4A5, Fairfax VA, U.S.A.
apirayes@gmu.edu, jpsousa@cs.gmu.edu

Abstract. Cell phones are assuming an increasing role in personal computing tasks, but cell phone security has not evolved in parallel with this new role. In the class of systems that leverage cell phones to facilitate access to remote services, compromising a phone may provide the means to compromise or abuse the remote services. To make matters concrete, SoonR, a representative off-the-shelf product is used to examine this class of systems from a security point of view. This paper identifies the shortcomings of existing solutions, and explores avenues to increase security without compromising usability. The usability of two proposed techniques is evaluated by means of a user study.

The contribution of this paper is a set of guidelines for improving the design of security solutions for remote access systems. Rather than proposing a one-size-fits-all solution, this work enables end-users to manage the tradeoff between security assurances and the corresponding overhead.

Keywords: Mobile computing, Remote PC access, Security, SoonR, User studies, Visual cryptography.

1 Introduction

In recent years, cell phones have evolved spectacularly from supporting telephony only to supporting multiple features, ranging from capturing and playing digital media, to e-mail access, to e-banking [1, 2], to remote access to personal files [3, 4].

Cell phone security, however, has not evolved at the same pace. There exist recent examples of high-end phones which support sophisticated security features. For example, NTT DoCoMo in collaboration with Panasonic produces a phone equipped with face-recognition and satellite tracking, and that automatically locks down when the user moves beyond a certain distance [5]. However, the vast majority of phones supporting the estimated 4 billion mobile service subscriptions in 2008 offer only the same password mechanisms used a decade ago [6].

With explosive popularity, their rising role in supporting daily activities of end-users, and with limited security mechanisms, cell phones are increasingly appealing targets for attackers.

Because today's phones are components of distributed software systems, a holistic view of security needs to be adopted: not only the phone and information it contains is at risk, but also is the information and services the phone has access to. Because an

J. Cordeiro et al. (Eds.): ICSOFT 2008, CCIS 47, pp. 76–90, 2009.
© Springer-Verlag Berlin Heidelberg 2009

attacker may obtain enough information to pose as the legitimate user to the remote services, physical protection of the phone is only part of the solution. These concerns are especially relevant for an emerging class of systems called *remote control applications* [7].

This paper focuses on a subclass of remote control applications where a cell phone facilitates access to files and services on a remote personal computer. To make matters concrete, we analyze and build on a commercial product, SoonR [3, 8], which is representative of this kind of solutions.

This work does not aim at improving the state of the art of security protocols, but rather at providing guidance for improving the engineering of solutions. For that, we analyze the threats and possible countermeasures for this class of applications.

A key principle for the work herein is to allow users to control the tradeoff between security and usability. Specifically, users may tailor the proposed security features to fit their needs: more concerned users incur more overhead in accessing services in a secure way, but reap greater assurances.

In the remainder of this paper, section 2 provides the background for remote access systems, and for SoonR in particular, while section 3 discusses the associated security issues. Section 4 proposes improvements to SoonR's current security model, Section 5 presents the design of a multi-factor authentication schema, and Section 6 discusses its implementation.

Section 7 presents the results of a study for evaluating the usability and security perceptions of end-users. Section 8 compares with related work, while Section 9 summarizes the main points in this paper and points at future work.

2 Background

New technologies such as RFID, GPS, Bluetooth, pointing and touching sensors, digital cameras, and image and voice recognition offer new opportunities to take cell phones beyond voice communication. Advertisement, tourist and museum guidance, electronic key, payment, peer-to-peer sharing, remote control, and field force tracking [9] are among these new applications.

Cell phone interaction with other devices can be categorized as short-range and long-range. The short-range interaction uses technologies such as Bluetooth, WiFi or USB. In contrast, the long-range interaction is mainly based on a communication over a computer network such as Internet. Although the short-range interaction can complement our work, the fundamental of this work is based on the long-range interaction via an IP-based network.

SoonR employs a Mobile Web 2.0 solution, which provides access to applications and files residing on a PC connected to Internet. Using SoonR, standard mobile phones capable of running a mini-web browser can use some applications on PCs remotely. For example, Google's Desktop Search, Outlook, Skype, and the files on a desktop computer become available at any location with cell phone network coverage. By using caching mechanisms, the files may be available even when the computer is turned off.

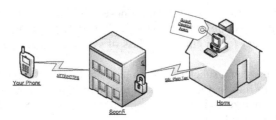

Fig. 1. Conceptual model of SoonR

For using SoonR, the user needs to download and run the SoonR Desktop Agent. As part of the configuration process, the user defines the folders to be shared and which applications the phone may access. Fig. 1 shows the key components of SoonR, according to SoonR's web site:

1. Cell Phone: with Internet access and web browser.
2. SoonR Desktop Agent (SDA): client software that links PC applications and data to the SoonR Service and reports changes to the computer's status.
3. SoonR Service: the broker between the cell phone and the SDA, the SoonR Service is responsible for authentication and storing meta-data of all shared resources.

3 Security Issues

Although SoonR supports authentication and encryption, several security concerns are left with no clear answers. Specifically:

1. Encrypted Communication: *optional* encryption is provided through SSL. According to SoonR's privacy statement, unencrypted access is also supported because older mobile phones still in use have no SSL support [8].
2. Device Authentication: the SoonR's registration process asks for a cell phone number, make and model. In addition, according to SoonR's privacy statement, when users login from their phone, the SoonR Software collects some information about the make and model of the user's mobile phone. However it is unclear what role this information plays in securing the service: in practice, users are allowed to access the service from a web browser running on any device.
3. User Authentication: authentication plays a key role in protecting against unauthorized access. However, given the limited keyboard available on cell phones, it becomes tempting for users to save the SoonR username and password on the mobile device. Unfortunately, this convenience may come at a high price if the phone is compromised. Locking the phone with a password is one possible countermeasure, although the numeric content and limited length of such passwords makes them easy targets of brute-force attacks.
4. Always-On: the SDA is always running, even when the user is not accessing the computer. Unfortunately, this always-on policy allows intruders plenty of opportunities to launch their attacks. Furthermore, the SDA doesn't show any warning if another browser is connected to the computer or if there is a flow of data between the personal computer and the server.

5. Trust in the SoonR Service: to optimize latency of remote access, user files are cached on SoonR servers, making these files are as secure as the servers themselves. However, SoonR's privacy statement states that while the company strives to use commercially acceptable means to protect users' information, absolute security cannot be guaranteed. No technical information is revealed about the means used to protect user data.
6. Trust in the SDA: by installing the SDA, the user trusts that it is doing what SoonR advertises it to do. Furthermore, an honest SDA may be compromised by a virus or through buffer overflow exploitation. In general, any program that opens a door to the internet may easily open a covert channel to transfer sensitive data in the background.

Table 1 summarizes these vulnerabilities and proposed improvements, which are elaborated in sections 4 and 5.

Table 1. Security threads and possible countermeasures

Security Issue	Proposed Solution
Always-On	Make it on-demand
Unauthorized use of phone	Authenticate the user
Using unauthorized device	Authenticate the cell phone
Saved SoonR password	Use one-time passwords
Caches on SoonR server	Clean up after each use
Trust on the SoonR agent	Monitor the agent's actions

4 Proposed Security Model

This section describes the assumptions and proposes a security model that improves on the model used by SoonR and other similar systems. Our work combines multi-factor authentication, one-time passwords, and device authentication to enhance intrusion deterrence. Additionally, to reduce exposure, we replace always-on by on-demand access.

The phone used in this work has capabilities commonly found in today's devices:

- voice with unblocked Caller ID,
- Internet connection with SSL-enabled browser, and
- Short text Messaging System (SMS) or Multimedia Messaging Service (MMS).

Although we believe that mobile service providers are trustworthy and do their best to secure voice and data channels, we assume no guaranties about the security of the cell phone device itself, or about malicious 3rd parties who may overhear the communications between the cell phone and the tower.

By the same token, and although the Caller-ID feature contributes to authenticate the device, a man-in-the-middle can fool the called device into displaying arbitrary Caller-ID information. Therefore we combine SMS (or MMS) with Caller-ID to authenticate the device.

Because not all users have the same security needs, our solution supports enabling or disabling each of the proposed security mechanisms. These mechanisms assume that the user's computer sits behind a well-configured firewall, runs an up-to-date operating system and anti-virus, and is physically accessible by authorized users only.

Furthermore, we assume that in addition to being connected to the internet, the user's computer is able to interact with a telephony system, such as VoIP software, or has a landline connected to a modem. The computer should be able to command the phone, pick up the line, play a message, receive a voice message and hang up. Also, the computer should be able to process the saved voice messages.

To address the limited keyboard capabilities available on cell phones, we experimented with speech technology to enhance the key-based interfaces provided by SoonR.

5 Improving the Design

In the current design, the SoonR Desktop Agent (SDA) is connected to the SoonR Service all the time. As discussed in section 3, it would be safer to establish a connection only when the user wants to use the service. However, at such time, the user is away from the computer and cannot start the SDA. In other words, on-demand connection is desirable from a security standpoint but is currently not supported.

This on-demand capability also should be password protected and capable of authenticating both the device and the user. A combined authentication of the user and the device eliminates the danger of unauthorized use of the cell phone.

The unauthorized use of a cell phone is a big threat to remote access solutions and it can happen at any time. As mentioned before, this becomes worse when users save their passwords on the cell phone. To overcome the unauthorized use of cell phone, which can be caused by lack of authentication on cell phone and the saved password, a simple one-time password is desirable. This one-time password can be an RSA token [10] or a simple shared random number.

In our secure scenario, users have the following installed on their computer:

– the SoonR Desktop Agent,
– a phone line with optional Caller ID capability connected to the computer, and
– the SoonR Watchdog, a new piece of software that implements the proposed security model and controls the behavior of the SDA.

Before connecting to their PCs via cell phone, users are required to make a regular call home. When this phone call takes place, the SoonR Watchdog detects the Caller ID, picks up the phone and prompts for a password using speech synthesis. If the user says the correct password, the SoonR Watchdog proceeds with authentication.

Because of the possibility of forging Caller-IDs, the Watchdog performs an additional authentication step using one-time passwords. Security tokens are an example of one-time passwords in practice [11]. While security token can provide a good level of security, installing and maintaining such system is not affordable for most home users. In addition, users need to always carry an extra piece of hardware. A cell phone-based token such as what has been presented in [12] would be an alternative replacement to the proprietary hardware tokens. Although this idea is appealing, it fails to protect the security and privacy of the user in case cell phone is lost or stolen.

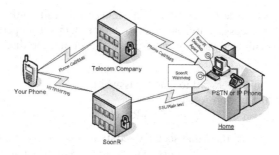

Fig. 2. Proposed extensions to SoonR's security

Furthermore, this solution requires a considerable effort for installation and maintenance.

The Watchdog implements a half-password solution, which, depending on the user's choice, consists of sending either an SMS containing a 4-digit random number to the user's phone, or a black and white image hiding a 4-digit random number.

In the case of sending an SMS, the Watchdog and the user share a simple, yet secret formula, which is applied to the half-password by both sides to generate the full password. We call this the *Secret Formula* method (more in section 6).

The alternative method, *Secret Image*, uses visual cryptography techniques to hide a random number [13]. The Watchdog initially generates a random image pad which is copied to the cell phone. For each authentication, the Watchdog picks a random number, writes it graphically on a fresh black and white image, and encrypts this image. The encrypted image is sent to the phone, and after combined with the initial pad, the visual of the random number can be recognized by a human.

Table 2 summarizes the one-time password mechanisms and their pros and cons for the purpose of the problem at hand.

Table 2. Comparison of different methods for generating a one-time password

Method	Pros	Cons
Security Token	Higher degree of security	Users carry extra item; cost; not suitable for home users
Cellular Authentication Token	User doesn't carry extra item; might be free	Doesn't help if cell phone is stolen; needs installation; not all mobile phone are capable of handling this method; different software for different mobile platform
Secret Image	Very low cost	Doesn't help if cell phone is stolen; not all cell phones are capable of handling this method
Secret Formula	Very low cost	User needs to remember the function; calculation may not be easy to all users;

Once a new password is generated and transmitted to the user by either method, the Watchdog interacts with the SDA to change the password on the SoonR service to the new password (see section 6). Fig. 2 shows the proposed security model.

After each access session, users should make a second call home to disconnect the service. This call is necessary for maintaining the on-demand property and without it we fall back to SoonR's always-on nature. After receiving the disconnect order, the SoonR Watchdog:

1. changes the SoonR password to a new password,
2. stops the SoonR Desktop Agent,
3. cleans up the server cache via the SoonR web interface

Since not every user has the same security requirements, our solution can be configured accordingly. To begin with, users may decide whether or not to use the SoonR Watchdog. If so, and except for the Caller-ID authentication, all other security features may be selectively deactivated, as explained in the following section.

6 Implementation

The SoonR Desktop Agent is available for MS-Windows and Mac operating systems. For practical considerations we implemented our solution over MS-Windows using Visual Studio 2005, although it would be possible to implement over Mac as well.

The system's interaction with the phone infrastructure uses the trial version of VTGO PC Soft phone, a product of IP blue Software Solutions [14] with Cisco CallManager 4.1 as a PBX. VTGO has an API [15] that exposes the phone functionalities to 3rd-party applications, allowing them to control the calls while VTGO is running in the background. Other possible choices for telephony would be SkypeIn [16] from Skype, and combination of Asterisk PBX [17] and IDefisk soft phone [18].

At present, SoonR does publish an API, and therefore we used AutoItX ActiveX Control as a workaround to interact with the SDA [19]. AutoItX is freeware for automating GUI interactions by replay [20]. The limitation of this workaround is that GUIs are more likely to change than published APIs, and any change in the SDA's GUI implies an update on the part of our experimental program that interacts with it.

Windows Speech Recognition technology powers the user voice interaction in our SoonR Watchdog [21]. We use Microsoft's Speech API, SAPI 6.1, to analyze the user's recorded message and detect the password. SAPI is freely redistributable and can be shipped with any Windows application that wishes to use speech technology.

We also use Windows XP Text-to-Speech capability to generate a dynamic phone conversation with user. First we generate an 8 kHz-16Bit-Mono '.wav' file and then we play it on the phone.

The Watchdog's voice recognition and text-to-speech behavior is controlled by Speech Properties applet of Windows' Control Panel. Our grammar is limited to digits 0-9 and "start/finish" commands. It is worth mentioning that we observed a significant improvement of recognition accuracy of version 6.1 of Microsoft Speech Recognition over version 5.1.

As discussed in section 5, the proposed authentication schema uses a one-time password mechanism. For the purpose of this work we came up with two ideas for one-time password: Secret Formula and Secret Image. In the Secret Formula method, the system and user both know a simple yet secret mathematical function, which should be applied to the shared password to generate the final password. The shared password is nothing but a random number. When the system verifies the voice password of the user, it picks a random number and sends it to the user via SMS. Then both parties apply the secret function to generate the new SoonR password. For example the secret function can be something like:

```
SoonR Password ="Pi" + ((Random-Number * 100) + 1) +"$"
```

Both user and the SoonR Watchdog calculate this new password. The result of this math function is the one-time password for the SoonR service, which will be set by the SoonR Watchdog and will be used by SoonR to authenticate the user. For example if the random number is 1234, the one-time password with this formula becomes Pi123401$.

This method is not very convenient and might not be as strong as hardware token but it is easy to implement and it doesn't need extra hardware of software to generate the password.

In the Secret Image method, we store the shared image pad on the cell phone in a place that it can be accessed by the cell phone mini-web browser or email client. The non-shared image is sent to the mobile phone as attachment to an email with a body similar to this:

```
<html>
 <style type="text/css">
 * {margin: 0; padding: 0;}
 </style>
 <body background="file://Photos/SecretImage0.png"
style="background-repeat:no-repeat;">
 <p><img src="cid:SecretImage1.png"></p>
 </html>
```

In cases where the cell phone email client does not support embedded images in the email, the non-shared image is uploaded to a publicly accessible URL and referenced in the email as:

```
<img src="http://publiclyAccessibleURL/SecretImage1.png">
```

6.1 Discussion

The proposed enhancements were implemented without accessing any API or the application's source code. Specifically, we proposed multi-factor authentication for the identities of both the user and the phone device.

One-time passwords contribute to overcome some of the security shortcomings of cell phones (see Table 2). Moreover, sending encrypted half-passwords to the phone, which both Secret Formula and Secrete Image do, protects against eavesdropping.

However, only Secret Formula prevents impersonation in the case of lost or stolen cell phone, since the generation of the full password relies on something the user knows: the formula. In the case of Secret Image, the user does not have to remember anything, but the fact that the image pad is stored on the phone makes it vulnerable to impersonation in case the device is lost or stolen.

Speech recognition technology is part of the proposed authentication process. Speech recognition is an exciting technology that promises to change the way people interact with computers. Speaking is the most widely used way of communication among humans. Considering the fact that in ubiquitous computing world, computers will be part of our daily life, it is natural that speech technologies become a significant player in the future of human-computer interactions.

Recent advances in speech recognition technology coupled with the advent of modern operating systems and high power affordable personal computers have culminated in the first speech recognition systems that can be deployed to a wide community of users. Our simple speech-enabled application is a good example of such systems. An end-user can use SoonR Watchdog without being bothered with technical details of speech recognition. During our experiment, we observed that training the speech recognition doesn't seem to affect the accuracy of recognition, possibly due to the extremely simple grammar. Ambient noise seems to be the determining factor.

7 User Study

We conducted a user study to evaluate the usability of the Secret Formula and Secret Image methods of authentication. For this study we used a mobile phone emulator onto which the shared secret image was loaded [22].

The participants were 38 graduate students of a user interface design class, with ages ranging from 20 to 36, and of which 15 were female and 23 male. The vast majority of the participants were professionals in the software industry who take classes after work. All the participants had a cell phone and 48% used their cell phone to browse the internet.

In this sample, males were ahead of females in taking cell phones beyond telephony. Males used cell phones for browsing the internet more often than females, 56.5% among all males to 33.33% among all females. Males also used their phone to access a remote PC more often: 17.4% among all males, to 6.7% among all females.

Not surprisingly, males used high-end cell phones such as Blackberry or iPhone more often than females: 30.4% among the males, to none among the females. All iPhone and Blackberry owners used the cell phone to browse Internet.

Although 89.5% of the participants said they would consider using remote access applications if they were easily available, only 16% do so with current technology. Both male and female participants stated a high level of interest in using remote access applications: 91.3% among all males, and 86.7% among all females.

There was also a gender difference concerning security. Overall, 71% of participants were concerned about the security of their phones, 91.2% among males and 60% among females, but only 50% of the participants used password for securing their cell phones: 78.2% among males and 26.7% among females.

After a quick explanation of the goals and principles of the study, participants were asked to register for Secret Formula. A simple web interface offered by the SoonR Watchdog enabled participants to build a formula with a pre-string, two basic mathematics operations and a post-string, all optional. Fig. 3b shows a screen shot of this interface for the formula exemplified in section 6. Participants were free to set a formula of their own choice.

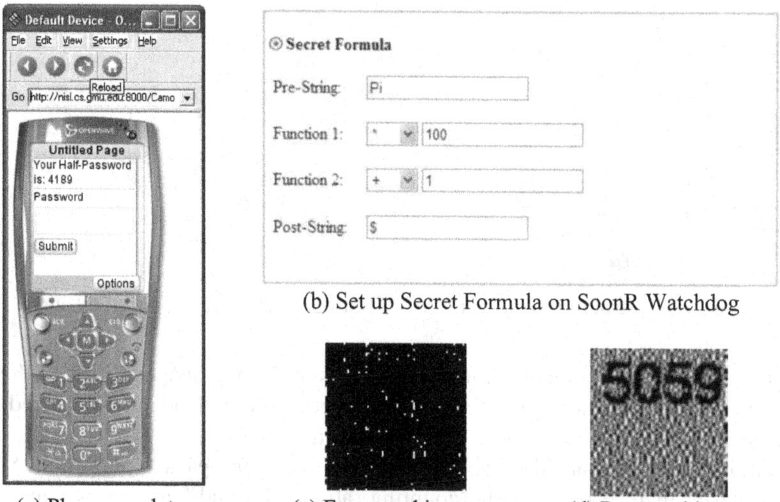

(a) Phone emulator

(b) Set up Secret Formula on SoonR Watchdog

(c) Encrypted image

(d) Password image

Fig. 3. Interfaces used for the user study

Participants were then asked to authenticate using Secret Formula first, as many times as they wanted, and then using Secret Image, also as many times as they wanted. Fig. 3a shows a screen shot of the phone emulator for authenticating using Secret Formula.

Fig. 3d shows an example of an image presented to the user for authentication with Secret Image, and which results from combining the initial pad with the half-password. As discussed in section 5, both the initial image pad stored on the cell phone and the half-password sent on the wire are encrypted images. Fig. 3c illustrates how an encrypted image looks like.

After experimenting with the interfaces for as long as they wanted, participants were asked to fill out a short survey. Fig. 4 shows the results of this survey concerning the ease of use of each authentication method (darker bars,) whether participants would consider using the method on a regular basis (lighter bars,) and whether participants would feel safer as a result of doing so (intermediate tone bars.) The answers to these questions were rated on a Likert scale, ranging from 1, strongly disagree, to 5, strongly agree.

Interestingly, although female participants were less concerned about security of their phone in general (see above,) they would consider using Secret Formula more often than males, 80.0% among females to 72.7% among males, considering a rating of 4 or 5 as a positive answer. Moreover, females would feel safer by using Secret Formula: 78.57% among females to 61.9% among males.

Among a total of 72 authentication attempts using Secret Formula, 44, or 61% were successful. Participants were asked to explain the failure, whenever it occurred, and all attributed it to the difficulty of calculating the full password based on the formula they had chosen. During the study, we observed that many participants resorted to pen and paper to aid them with the calculations, although admittedly leaving such notes behind would constitute a serious breach of the formula's secrecy.

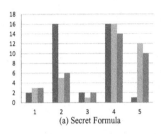 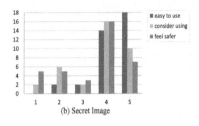

(a) Secret Formula   (b) Secret Image

Fig. 4. Survey responses on a Likert scale: 1-strongly disagree to 5-strongly agree

Among a total of 49 authentication attempts using Secret Image, 43, or 88% were successful. Participants were asked to explain the failure, whenever it occurred, and all attributed it to difficulty in using the emulator interface.

When asked to compare the two methods, 91.7% of the participants think Secret Image is easier than secret formula (counting ratings of 4 and 5 as positive answers.) Moreover, 61% of the participants indicated an overall preference of Secret Image over Secret Formula. The difference between these two numbers reflects the fact that security assurances was considered by the participants as an important factor, in addition to ease of use. In fact, participants are widely divided over the strength of the two methods. 33% agree or strongly agree that Secret Image offers at least as much protection as Secret Formula, and 20% have no opinion. 47% of the participants were aware that Secret Image actually offers less protection than Secret Formula in the case the phone is lost or stolen.

8 Related Work

In addition to SoonR, there exist other commercial solutions with similar security issues. For example, PocketView, which is GoToMyPC's handheld version. As with SoonR, the host PC initial communications are established through a TCP connection brokered by a provider's server. Because there is an assumption that the outgoing connections generally do not pose a security risk, most firewalls allow them without any configuration required and after establishing the channel, the connection remains open for virtually an unlimited time.

SoonR and PocketView are not the only examples. In early 2005, Toshiba announced the world's first software supporting remote operation of a personal computer from a mobile phone [23]. According to Toshiba's press release, the Ubiquitous Viewer software provides access to any Windows home or office computer. It also allows users to open productivity software, such as the MS-Office suite, PC-based e-mail, Internet browser and other PC applications at any time, wherever they are. Unfortunately there is no public information about this application so we were not able to judge the security of the system.

Makoto Su et al. propose a simple solution for using the joystick available in some phones to control a cursor and perform clicks on a remote display, but does not address security issues [24].

In other work, a user at an distrusted terminal utilizes a trusted device such as PDA to access the home PC securely through by remote desktop application such as Go-ToMyPC [25]. The PDA as input device and distrusted terminal as display, create a secure environment that a possible Spyware on distrusted terminal cannot capture the user input on the distrusted terminal. [26] introduces a prototype of a proxy-based system in which the user must possess a trusted mobile device with communication capability such as a cell phone, in order to securely browse the net from distrusted terminal. In this system, the trusted proxy contacts the user through the trusted device to obtain authorization for requests from the un-trusted computer.

All the mentioned products, systems and projects assume the handheld devices are secure enough to be utilized for accessing other devices. However this assumption is far from reality. What makes our work different is that we don't trust the device and its holder. Instead of blindfolded trust, we try to authenticate the mobile device and the device holder to the best of our ability.

The closest to our work is [27] where the authors employ home-automation techniques to grant properly authenticated users, remote access to the their personal workstations at home. The solution uses web interface, short message text and phone call to authenticate the user. For example in short text message authentication schema, the user composes an SMS message by concatenating his account, password, the name of the personal workstation to be controlled remotely, and the interface address of the public terminal, in comma separated format. Then he sends the message to the phone number associated with his residential gateway. Upon receiving the message, the residential gateway parses it and performs validation on the authentication information. If nothing goes wrong, the residential gateway turns on the specified personal workstation, set up the temporary NAT and PATS port-mapping entries, and sends back another SMS message containing the corresponding URL.

The main difference between our system and Tsai et al. is that we never assume the SMS, MMS or phone calls are secure. With our one-time password mechanism the danger of eavesdropping is minimal.

Visual cryptography was introduced by Naor and Shamir at EUROCRYPT '94 [13]. The algorithm specifies how to encode pixels of an image into two or more shares in a way that an image can be reconstructed "visually" by putting shares on top of each other. [28] and [29] explain examples of visual cryptography applications. The main difference of our work and previous works is that while they emphasize on "very common low-tech technology" applications such as transparencies we use simple techniques of HTML to put the shares on top of each other.

9 Conclusions

This paper demonstrates that it is feasible to improve the security assurances of a commercial off-the-shelf product. The product used in this case study, SoonR, is representative of an emerging class of commercial products for accessing remote PCs using a cell phone. Specifically, the proposed enhancements consist of:

- Reducing the window of exposure to threats by granting remote access to the user's PC only when required, instead of supporting the current always-on policy.

- Reducing the likelihood of impersonation by using multifactor authentication, specifically:

    a) Verifying the phone's caller id,
    b) Asking a one-time password from the user,

- Reducing the vulnerabilities in case the cell phones are lost or stolen by having the one-time password being generated by "something the user knows," along with "something the user have."

An important feature of the proposed solution is that it enables users to manage the tradeoff between security assurances and the associated usability overhead. Users with stringent requirements may use sophisticated mechanisms for authentication, such as generating complex one-time passwords combining "something I know," with "something I have." Users who value the ease of access more may rely on simpler mechanisms, such as visual cryptography, which are based on "something I have." Pushing ease of use even further, users may skip one-time passwords altogether and rely on device caller id alone.

With respect to evaluation, we conducted a user study which addressed both the usability of the proposed solutions, and users' current habits and perceptions concerning remote access applications in general. In a sample of graduate students, a very significant majority looks positively on both remote access applications, and on the proposed mechanisms for making them more secure. Although there is room for improvement, the proposed authentication methods proved to be usable with a moderate to high success rate.

In future work, we will try to improve the usability of one-time passwords, namely the Secret Formula method, possibly by incorporating ideas from visual cryptography.

Although the mechanisms described in this paper considerably improve the security aspects of SoonR, further improvements are possible. Monitoring the system calls made by the SoonR Desktop Agent would offer protection against malicious or vulnerable code. For example, a compromised agent might try to access areas of the hard disk which are not defined in the configuration file. Additional security practices could be added to the Watchdog, such as event logging, password aging, strong password policy and maximum number of failed authentication attempts.

References

1. Hamilton, A.: Banking Goes Mobile. TIME Magazine (2007),
   http://www.time.com/time/business/article/
   0,8599,1605781,00.html
2. Tiwari, R., Buse, S., Herstatt, C.: Mobile Services in Banking Sector: The Role of Innovative Business Solutions in Generating Competitive Advantage. In: International Research Conference on Quality, Innovation and Knowledge Management, New Delhi, pp. 886–894 (2007)
3. SoonR: SoonR - In Touch Now, The Company, http://www.soonr.com
4. Kallender, P.: Toshiba Software will Remotely Control PCs by Cell Phone. Computer World (2005),
   http://www.computerworld.com/softwaretopics/software/story/
   0,10801,99012,10800.html

5. Kageyama, Y.: Cell Phone Takes Security to New Heights. The Associated Press (2006)
6. UN News Center: Number of cell phone subscribers to hit 4 billion this year, http://www.un.org/apps/news/story.asp?NewsID=28251&Cr=Teleco mmunication&Cr1
7. Roduner, C., Langheinrich, M., Floerkemeier, C., Schwarzentrub, B.: Operating Appliances with Mobile Phones - Strengths and Limits of a Universal Interaction Device. In: LaMarca, A., Langheinrich, M., Truong, K.N. (eds.) Pervasive 2007. LNCS, vol. 4480, pp. 198–215. Springer, Heidelberg (2007)
8. SoonR-Privacy-Officer: Privacy Policy (2007), http://www.soonr.com/web/front/security.jsp
9. Enrico, R., Wetzstein, S., Schmidt, A.: A Framework for Mobile Interactions with the Physical World. In: Proceedings of the Wireless Personal Multimedia Communication Conference (WPMC 2005), Aalborg, Denmark (2005)
10. RSA SceurID. http://www.rsa.com/node.aspx?id=1156
11. Security token. Wikipedia, http://en.wikipedia.org/wiki/Security_token
12. Di Pietro, R., Me, G., Strangio, M.A.: A two-factor mobile authentication scheme for secure financial transactions. In: International Conference on Mobile Business, pp. 28–34 (2005)
13. Naor, M., Shamir, A.: Visual Cryptography. In: De Santis, A. (ed.) EUROCRYPT 1994. LNCS, vol. 950, pp. 1–12. Springer, Heidelberg (1995)
14. IP blue Software Solutions, http://www.ipblue.com/
15. VTGO API - Programmer's Reference Guide, http://www.ipblue.com/documents/ VTGO-API-Programmers-Reference.pdf
16. SkypeIn - your personal number. Skype, http://www.skype.com/allfeatures/onlinenumber/
17. Van Meggelen, J., Smith, J., Madsen, L.: Asterisk: The Future of Telephony. O'Reilly Media, Inc., Sebastopol (2005)
18. idefisk, softphone, http://www.asteriskguru.com/idefisk/
19. Autoit, http://www.autoitscript.com/autoit3/
20. Flesner, A.: AutoIt v3: Your Quick Guide O'Reilly Media (2007)
21. Speech API SDK, Microsoft, http://www.microsoft.com/speech/techinfo/apioverview/
22. Openwave: Openwave Phone Simulator, http://developer.openwave.com/dvl/tools_and_sdk/ phone_simulator/
23. Kallender, P.: Toshiba software will remotely control PCs by cell phone. COMPUTER WORLD Today's top stories (2005), http://www.computerworld.com/softwaretopics/software/story/ 0,10801,99012,10800.html
24. Makoto Su, N., Sakane, Y., Tsukamoto, M., Nishio, S.: Remote PC GUI Operations via Constricted Mobile Interfaces. In: 8th Intl. Conf. on Mobile Computing and Networking, pp. 251–262. ACM Press, Atlanta (2002)
25. Oprea, A., Balfanz, D., Durfee, G., Smetters, D.: Securing a remote terminal application with a mobile trusted device. In: 20th Conf. on Computer Security Applications, pp. 438–447 (2004)
26. Jammalamadaka, R.C., van der Horst, T.W., Sharad, M., Seamons, K.E., Venkasubramanian, N.: Delegate: A Proxy Based Architecture for Secure Website Access from an Untrusted Machine. In: 22nd Conference on Computer Security Applications, pp. 57–66 (2006)

27. Tsai, P., Lei, C., Wang, W.: A Remote Control Scheme for Ubiquitous Personal Computing. In: IEEE International Conference on Networking, Sensing & Control, Taipei, Taiwan (2004)
28. Ateniese, G., Blundo, C., De Santis, A., Stinson, D.R.: Visual Cryptography for General Access Structures. Information and Computation 129, 86–106 (1996)
29. Naor, M., Pinkas, B.: Visual Authentication and Identification. In: Kaliski Jr., B.S. (ed.) CRYPTO 1997. LNCS, vol. 1294, pp. 322–336. Springer, Heidelberg (1997)

Understanding and Evaluating Replication in Service Oriented Multi-tier Architectures

Michael Ameling[1], Marcus Roy[1], and Bettina Kemme[2]

[1] SAP Research CEC Dresden, Chemnitzer Str. 48, Dresden, Germany
{michael.ameling,marcus.roy}@sap.com
[2] School of Computer Science, McGill University, 3480 University Street, Montreal, Canada
kemme@cs.mcgill.ca

Abstract. Current service-oriented solutions use a multi-tier architecture as their main building block. These architectures have stringent requirements in terms of performance and reliability. Replication is a general solution to provide fast local access and high availability. At the application server tier this means replicating the reusable software component of the business logic and the application dependent state of business data. However, while replication of databases is a well explored area and the implications of replica maintenance are well understood, this is not the case for data replication in application servers where entire business objects are replicated, Web Service interfaces are provided, main memory access is much more prevalent, and which have a database server as a backend tier. In this paper, we introduce possible replication architectures for multi-tier architectures. Furthermore, we identify the parameters influencing the performance of the replication process. We present a simulation prototype that is suitable to integrate and compare several replication solutions. We describe in detail one solution that seems to be the most promising in a wide-area setting.

1 Introduction

Multi-tier architectures are the main building block of modern business applications. Clients submit requests to application server tier which implements the application logic (e.g., purchase orders, bookings, etc.). The persistent data is stored in a backend tier.

Multi-tier platforms often host many applications at the same time which might interact with each other, or with applications hosted on other platforms. As businesses open their applications to a wide range of clients the infrastructure is put on stringent requirements. Firstly, it has to follow a service-oriented architecture (SOA) where the business logic is exposed as services through well defined interfaces. Secondly, the platform must be able to support an increasingly heavy and complex load. Thirdly, it has to provide a certain QoS level not only to local clients but to clients across the globe. Finally, the platform must be available around the clock. One of the main techniques to achieve these objectives is replication. It allows the system to distribute the load, to provide fast access to close-by replicas and to increase availability as clients can connect to any of the available replicas. A main challenge of replication in data intensive environments is replica control, that is, the task to keep the data copies consistent.

Replication has been widely used in database systems and the implications of using replication are well understood [1,2,3,4]. Similarly, object-based systems have explored

J. Cordeiro et al. (Eds.): ICSOFT 2008, CCIS 47, pp. 91–104, 2009.

replication as a tool to either increase fault-tolerance [5,6,7] or improve performance [8]. However, only recently replicating the middle-tier of a complex multi-tier system has been considered [9,5,10,11,12,13]. So far, the proposed solutions have been developed for very specific scenarios, and it has become clear that no single solution will fit all. The reason is that the impact of replication on performance depends on many different parameters such as the required degree of consistency, the distribution within LANs or WANs, the workload, etc. As implementing replica control into a complex multi-tier architecture is a non-trivial task, one has to understand the implications of the different replication alternatives in order to choose the best solution for the given scenario.

In this paper we present a step by step a framework that allows reasoning about replication alternatives. We first discuss the many different ways applications are deployed within a SOA, and how replication can be deployed in these environments (Section 2). Secondly, we discuss a range of replication solutions developed within the database community, and discuss their suitability for SOA (Section 3). Thirdly, we have developed a simulation framework that captures the most significant characteristics of SOA (Section 4). Thus, it serves as a first analysis platform and provides a means to compare different replication alternatives before any real, and thus complex and cumbersome implementation is done. We show an initial evaluation using our simulator (Section 5).

2   Replication in SOA

2.1   Basic Architecture

A business application maintains a set of *business objects*, and defines a set of *services* as its interface. Clients can call these services to create, delete, access and manipulate the business objects. In order to provide interoperability and support heterogeneous environments, the interfaces usually follow the web-service standard. Examples of applications are "customer relationship management (CRM)", "supply chain management (SCM)" or "project management (PM)". A typical example of a service within a CRM is "customer data management" providing a set of methods that allow the customer to change customer related information such as the address. A business object in this context would assemble all information related to the customer.

Figure 1 shows the architecture in which these applications are embedded. An *application server (AS)* is a software engine hosting one or more applications. It provides the computing environment to control and schedule service executions on the applications. The applications hosted on an AS can belong to the same or different companies (e.g., in the case of data centers). In any case, the usage of AS allows for an efficient sharing of resources. In standard computing environments, an AS can host up to 20 different applications each serving up to 100 clients concurrently.

Clients connect to one of the applications, and then can call the services offered by this application. If a client wants to access more than one application, it typically has to connect to each individually. AS and backend tier, typically a database system (DBS) are usually located close to each other, often even on the same machine. AS and DBS share the responsibility of maintaining the state of the business objects. The DBS provides the persistence of business objects. However, during service execution, the AS loads the objects from the DBS into its main memory cache. While the database

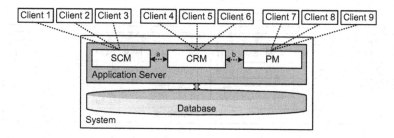

Fig. 1. Multi-tier Architecture

maintains the data in form of records of tables, the AS provides an object-oriented view. Often, a business object is assembled from many different database records providing a more appropriate abstraction for application programming. Changes on business objects are written back to the database when service execution completes.

Most commonly, each application has its own distinct business objects, and maintains its own database (or at least its own tables) within the DBS. This allows for a clear separation of components and is necessary to guarantee reusability. In order to exchange information between the applications, they have to call each others' services through the standard interfaces making one application the client of the other. That is, the service execution within one application might call a service of another application.

2.2 Replication

Replication can boost performance in two ways. In *cluster replication*, server replicas are used for scalability. A load-balancer component distributes incoming service requests to the replicas. As more replicas are added, the system is able to handle more load. Alternatively, replicas can be located at distant geographic locations. Thus, clients can connect to the closest replica, avoiding high-latency WAN communication between client and server. Also, the system remains available even if remote servers (e.g., head quarters) are currently not accessible.

As SOA environments maintain a considerable amount of data, we cannot simply replicate server instances – we also have to replicate data. As data can change, this requires some form of *replica control* whose task it is to keep the data copies consistent. Since the maintenance of data is distributed across AS and DBS the question arises who is responsible for replica control. There exists a large body of database replication solutions, making it attractive to rely on replica control at the DBS level. However, this does not seem practical in SOA because it either affects efficiency or consistency. The business objects maintained by the AS layer can be viewed as a cache of the data stored at the DBS. If only the DBS layer is responsible for keeping data copies consistent, each access to a business object has to be redirected to the DBS in order to guarantee to read the latest version of the data or to make sure that all copies are updated. With this, the efficiency advantages of the AS object cache are lost. Alternatively, one could choose to use the cache anyways instead of accessing the database. However, then, no guarantees could be given in regard to the freshness or the consistency of the data.

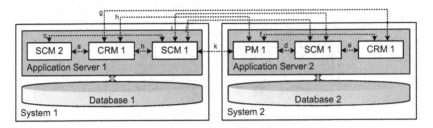

Fig. 2. Replication in SOA

Therefore, we envision an architecture, where the AS layer is responsible for replica control (Figure 2). There are several AS instances, and each is connected to its own DBS which can be off-the-shelve software without replication semantics. Each application can run on several AS instances. The objects associated with an application are replicated across the AS instances. For simplicity we assume full replication, i.e., each object has a copy at each application instance. The AS layer performs replica control. A replica control mechanism at the AS layer is responsible for keeping the object copies consistent.

In principle, there does not seem to be a difference in architecture if the connection between the AS instances is a LAN or a WAN. However, the fundamental latency difference between LANs and WANs and the different purposes for which cluster replication and WAN replication are applied, require very different replica control solutions and we will discuss them shortly. In practice, large SOA are likely to integrate both cluster and WAN replication. At each location, a cluster solution distributes the load locally submitted. The different locations are then connected via a WAN replication solution.

3   Replication Strategies

The most crucial overhead is replica control, i.e, the task to keep the data copies consistent when updates occur. Replica control has been well explored in the database community and a wide range of replica control solutions exist that have fundamental influence on non-functional properties such as availability, data consistency, reliability as well as system performance. Although the replica control algorithms themselves are not directly applicable to AS replication, the design space is similar at both layers. Nevertheless, the many particularities of replication at the application layer have an influence of the design choices and have to be understood.

3.1   Categorization

Wiesman et al. [14] categorizes replica control algorithms by four parameters, partially taken from [1], developing a replication hierarchy (Figure 3).

Architecture. In a *primary copy approach*, each data item has a primary copy. Updates are always first executed on the primary copy which is responsible of propagating any changes to the secondary copies. Read-only access can be performed at any copy. A

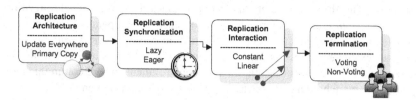

Fig. 3. Overview of Replication Hierarchy

typical setting is to have one replica owning all primary copies. This makes concurrency control easier: this primary site determines the serialization order and all others apply changes according to this order. However, the site can become a bottleneck. Furthermore, in a WAN setting, all update transactions, independently of where they are submitted, must go to the one primary – which will be remote in many cases, leading to long response times. Thus, if the data can be partitioned geographically, meaning each partition is accessed more frequently by clients of a certain region, then each replica can have the primary copies of the partition that is accessed locally, and secondary copies of the other regions. In contrast to primary copy, in an *update everywhere* approach, each site accepts update transactions on any data copies and propagates updates to other sites. This makes the approach more flexible since clients can submit their requests anywhere. However concurrency control requires now complex coordination among the replicas.

In the context of AS replication, an additional challenge is that the different AS instances might cache different data (the most recently used) while the remaining data only resides in the database. This makes coordination more difficult. In the AS context, a primary copy approach where each AS instance has the primary copies of a subset of the data, might be of advantage, as the caches of the different AS instances will then likely cache this data partition taking better advantage of the available cache.

Synchronization. *Eager replication* synchronizes all replicas within the context of the transactions. There are no stale replicas. However, communication and traffic complexity is high. In contrast, in *lazy replication*, a transaction with all its updates is first executed at one site, and then the changes are propagated to other replicas in a separate transaction. Transaction response time is shorter but the data copies are not always consistent. While eager replication is, in principle, more desirable than lazy replication, the potential response time increase might not be acceptable in WAN environments. Also, eager replication, if not designed carefully, might more easily lead to unavailabilities. However, eager replication is likely to be feasible in cluster replication.

In the context of AS replication, a further influence in the synchronization overhead is the fact that before data can be written to the database, it has to be transformed from the object-oriented format to the relational format. Thus, the question arises, whether eager replication only includes the transfer of updates to the other replicas without actually writing and committing the changes to the database, or whether it means the transactions have to be committed at all replicas before the user receives the response. The latter will mean, that eager replication in the AS layer will be even more expensive than at the database layer.

Interaction. The third parameter *interaction* decides how often synchronization is done. In a *linear interaction* approach, typically each write operation within a transaction leads to a communication step. In contrast, in *constant interaction* the number of messages exchanged does not depend on the number of operations in a transaction. While constant interaction is essential in WAN environments where communication is likely to be the main bottleneck, it is less crucial in LAN environments.

Again, AS replication puts extra overhead to the interaction. In database replication, propagating a write operation typically means sending the changes that were performed by the write operation. For that the database system has to keep track which attributes of which records have been changed. This is already done for logging purposes. Thus, sending the changes can be as easy as sending the log information about the updated records. In AS replication, however, tracking changes performed on the objects is much harder. Simple solutions perform object serialization which can have a huge overhead. Detecting the low-level changes performed by a write operation is non-trivial. An alternative solution would not send data changes but the write operation command and execute it at all replicas. However, this has its own overhead at the remote replicas. Additionally, it is problematic as it requires deterministic execution.

Termination. The final parameter decides on whether a transaction is terminated via *voting* or *non-voting*. Voting requires an additional exchange of messages in order to guarantee that all replicas either commit or abort the transaction.

Using the above four parameters is not the only way to classify replication solutions. In [15] the authors describe transaction execution with the help of five phases. Not all protocols have all phases while other protocols might execute a phase several times for each operation of a transaction. The client submits the transaction (or an operation of a transaction) to one or more servers in phase 1. In phase 2, the servers coordinate before execution, e.g., to decide on a serialization order. In phase 3 the transaction (or operation) is executed (at one, some or all servers). In phase 4 the servers agree on the outcome of the execution (it might also include the execution at servers that haven't done so in phase 3). Finally, in phase 5 the client is notified about the outcome. In lazy approaches, no phase 2 takes place, and phases 4 and 5 are switched as the client is first informed about the outcome and then the changes are propagated to the other servers.

The above description shows that AS replication solutions can be categorized in a similar way as this has been done for database replication protocols, and we can use the categorization to be aware of the solution spectrum that is possible. However, as our discussion above has already shown, two things have to be considered. First, one has to be aware that the protocols have to be adjusted to work at the middle-tier layer. In particular, the object-oriented view, the volatile state of data at the application server, and the interactions with a database backend-tier have to be taken into account. Secondly, we cannot simply assume that the general performance of the different algorithms will be the same as if they were implemented at the database level, given the particularities of AS replication.

3.2 Example Algorithm

We now describe one AS replication protocol in detail. We later use this algorithm to explain how an algorithm can be integrated into the simulator framework. Our algorithm is designed for a WAN environment where communication latencies are high.

Lazy Primary Copy. The algorithm we describe is lazy. This seems the most attractive in a WAN to avoid long client response times. We choose a primary copy approach since it simplifies concurrency control. Concurrency control at the AS layer is not yet well explored, doing it in a distributed and replicated setting would be even more challenging. Furthermore, the inconsistencies that can occur in a lazy environment if two different AS instances update the same item concurrently can quickly become a nightmare. A lazy approach makes constant interaction a natural choice because all updates are known at commit time, and thus, before propagation is initiated. Furthermore, voting can be avoided if the serialization order for each transaction can be determined by a single site. This is possible if all data items updated by this transaction have their primary copies at the same site.

As we mentioned above, lazy replication provides only then fast response times for updates if the submitting client is close to the primary site. In order to keep the number of remote accesses low, the primary copy of a data item should be located at a site that is close to most of the clients that want to update this data item. For instance, if the data item refers to a specific client, then the primary copy should reside on the AS that is close to the client's home location.

Replication on Object Level. The natural granularity for replication in an AS environment is a business object (BO) because it builds a logical unit within the business semantics. Note that this might be quite different to the logical unit in the database, i.e., a database record. Therefore we refer to our approach as *primary copy on the object level*. Figure 4 shows an example setup. There are four AS instances with their local DBS connected via a WAN. There are four business objects with primary copies A, B, C, D, and the respective secondary copies A', B', C' (D does not have a secondary copy). Each of the four AS instances is the primary for one of the BOs. The assignment of BO can be based on various conditions e.g., semantical or geographical. According to our experience, geographical affinity is often provided. We assume that each transaction only updates BOs for which a single site has the primary copies.

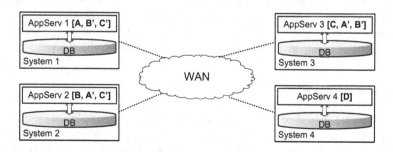

Fig. 4. Primary Copy Replication on Object Level

Protocol Description. We assume that each client request triggers a transaction within the AS. The transaction can consist of several read and write operations accessing local BO copies. The protocol proceeds as follows. When a client submits a read-only transaction to an AS instance, the local instance executes the request locally and returns to the client. No communication with other AS instances is needed. If the client submits an update transaction, e.g., updating A in the figure, if the AS instance has the primary copies of the BOs updated by the transaction (i.e., if it is system 1), the transaction executes and commits locally. The AS instance returns the result to the client and then forwards the update on the BOs to the secondary copies. If an update transaction is forwarded to a secondary copy (e.g., system 3), the secondary could either reject the transaction or transparently redirect it to the primary. The primary executes it and then forwards the changes and the response to the secondary (who sends them to the client) and all other secondaries.

The details of what exactly is forwarded (e.g., entire object, only changes, or the operation which then needs to be reexecuted at the secondaries), and how this forwarding takes place (via special communication mechanism or via web-services itself), might have a big impact on the performance. The simulation framework that we present in the next section is flexible enough to model such differences.

4   Simulation

A simulation, compared to a real implementation, has several advantages. Firstly, implementation is fast and does not require a real distributed environment as testbed. In contrast, implementing even a prototype replication tool into a real SOA is very complex, making it hard to implement several algorithms within a reasonable time frame. Secondly, prototype implementations are often not optimized, leading to bottlenecks and overheads that would not occur in any well-engineered implementation. A simulation can abstract from such artifacts. Finally, a simulation framework allows for a greater variety of parameters, and thus, a more versatile evaluation comparison. Therefore, we have designed a simulation architecture that captures the most important aspects of SOA, in particular in regard to performance implications. We show along one example how an algorithm can be implemented into our simulation framework, and how performance results can be derived showing the influence of various parameters. Our aim is to build a highly parameterized simulation framework for AS replication. It allows for an easy plug-in of various replication algorithms. It enables to analyze different hardware infrastructures, and it takes the particularity of a multi-tier architecture into account. For example, it allows to model replication at the object level.

4.1   Simulation Architecture

Our framework is based on J-Sim [16], a component-based simulation environment. J-Sim offers the concept of a *core service layer* (CSL) which implements already the basic services of a network layer. On top of this the *upper protocol layer* (UPL) provides the infrastructure for the replica control algorithms and emulates the standard AS

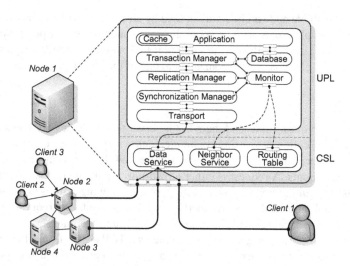

Fig. 5. Node Architecture

components. Our solution maintains the component-based, loosely coupled architecture of J-Sim, and thus, is easily customizable and extensible. Figure 5 illustrates the architecture of one individual node (AS instance plus database).

Components of UPL. The `Transaction Manager` component is responsible for transaction management. The `Application` component simulates the business logic and maintains the BOs. It has its own `Cache`. BOs can be created, read, updated and deleted. The AS is also connected to the `Database` for the persistence of BOs. These three components are also found in a non-replicated multi-tier architecture. The other components implement the replication functionality. They basically intercept requests from clients and responses sent to clients in order to control the replication process. The `Transport` component, implements the transport protocol, e.g., TCP. It is connected to the `Data Service` provided by the CSL of J-Sim. The next higher layer is the `Synchronization Manager` which is responsible to handle the propagation of replication messages. The `Replication Manager` component implements the main tasks and execution flow of the replica control algorithm, e.g., the lazy, primary copy scheme. The `Monitor` watches all components for monitoring reasons. It also contains concepts that are needed by all other replication components, e.g., routing tables. The architecture resembles the interceptor-based approach used to plug-in functionality into web- and application servers.

Resource Consumption. In the simulation setting, all components require CPU to perform their tasks. In order to measure and estimate the CPU usage, every component is additionally equipped with a time module. Each time the component is triggered to do something, the time module collects the time of processing and relatively computes a percentage that mirrors the CPU usage. That is, we do not simulate time but measure the actual time the execution requires.

Topology. The above node architecture is used by all servers (both primary and secondaries in the example protocol). The nodes are connected through the port connected to the `Data Service`. Clients use the same port to send their requests. (e.g., `Client 1` to `Node 1`). A client is always connected to one of the servers.

4.2 Execution

Clients generate requests and send them to the server to which they are connected. We show here the execution flow using the implemented lazy, primary copy approach on object level. Other protocols will follow a similar flow but perform different actions at different time-ponts. A request is either a read-only or update transaction accessing several BOs. The submission of a request triggers the following actions in the simulator.

Let's first look at a read-only transaction. A client submits the request in form of a byte stream to its local server. The request goes through the data service and is forwarded to the transport manager which implements a TCP abstraction. Thus, it models sufficiently well a web service request over a standard network and transport layer. The transport layer transforms the byte stream back into a request. The request goes through the synchronization mechanism (which usually does not have anything to do for client requests). Then it goes to the replication manager (nothing has to be done for a read request). When the request arrives at the transaction manager, a new transaction is started and the operations are submitted to the application layer which simulates some execution. If the requested BOs are not in the cache, they have to be loaded from the database, otherwise they can directly be read from the cache. After completion, the response moves back through the different layers. The transaction manager commits the transaction, the other layers don't have to do anything for a read request, and the client receives the answer.

Let's now discuss the steps when the client submits an update transaction directly to the server that contains the primary copies. As the request moves up the layers, nothing has to be done by the synchronization and replication managers since the approach is lazy and replication tasks are triggered after commit (eager replication might already perform some actions at this time point). The transaction manager again starts a transaction and the application simulates the processing of the write request. At the latest at transaction commit time the changes to the BOs are written to the database. When the result is returned to the replication manager, it detects a change state. It receives via the monitor all secondaries of the updated BOs (this meta-information is maintained by the monitor). It initiates via the synchronization manager the propagation of the state changes to these secondaries. At the same time, the result is also returned to the client. Secondaries receive the propagation messages also through the data service. The synchronization manager at the secondary determines that this is a propagation message (and no client request). An appropriate refresh transaction is started at the secondary to apply the changes on the BO copies.

In case the client submits a write request to a server that does not have the primary copies, the replication manager catches this request and it is forwarded to the primary which executes it as above. By letting clients and server both use the data service and then direct requests through the same layers, a variety of message types and control flows can be supported. When the secondary receives the refresh transaction from the

Table 1. Overview of Parameters

Domain	Parameter
System	N_S: Number of Servers, N_C: Number of simultaneous Clients N_O: Total Number of Objects, R_F: Replication Factor
Network	B_W/B_L: Bandwidth WAN/LAN, L_W/L_L: Latency WAN/LAN
Transport	P: Packet Size, MTU: Maximum Transmission Unit MSS: Maximum Segment Size
Transaction	C_I Interval of Client Transactions, $T_{R/W}$: Read/Write Ratio N_O: Number of Operations, $T_{/C}$: Transaction per Client
Database	D_R: Delay Read, D_W: Delay Write
Replication	R_A: Replication Algorithm

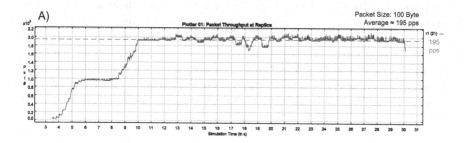

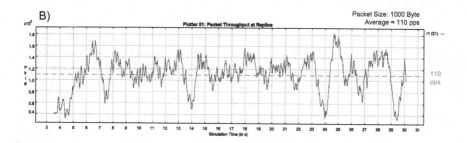

Fig. 6. Packet Throughput

primary, the synchronization manager sends the response to the client and at the same time forwards the update upwards so that it is applied on the BO copies and in the database.

4.3 Parameters

The behavior of an algorithm is influenced by many parameters. Our simulation framework should help evaluating the system under varying conditions and understand the implications of each parameter. So far, we focus on the impact on performance based on the parameters depicted in Table 1.[1] N_S is the number of servers in the system. Each server can have a different number of clients, thus, N_C is actually a list of size N_S, indicating the number of clients for each server. The total number of BOs in the system is N_O while R_F depicts the replication factor, i.e., the number of copies per object. If $R_F = N_S$ than all objects are replicated at all servers (full replication), otherwise the copies are equally distributed across the servers.

We distinguish the bandwidth and latency of WAN (B_W, L_W) and LAN (B_L, B_W) messages. The system takes as input a topology graph that indicates whether links are WAN or LAN. For instance, all client/server links could be LAN while all server/server links are WAN. The transport layer has the typical TCP related parameters. For transactions, the overall load in the system is determined by the time C_I between two consecutive transactions submitted by an individual client. Furthermore, each transaction has a number of operations N_O and a ratio $T_{R/W}$ of read operations among all its operations. Each operation accesses one of the BOs randomly. When executing an operation in the application, each read and write causes a delay for the database access (D_R and D_W).

If behavior over time should be measured, one can indicate $T_{/C}$ as the number of transactions per client until the simulation terminates. If averages should be calculated (e.g., average response time), the simulation runs until a confidence interval is achieved. Finally, the chosen replication algorithm is an important parameter influencing the performance.

5 Selected Results

In this section, we shortly show how our simulator can be used to analyze the performance using the lazy, primary copy implementation. Figure 6 shows the impact of packet size assembled for transmission over the network. The experiment shows how a rather low level parameter can have a tremendous effect on performance. There exist one primary and one secondary server. Each server has a single client. Every client submits a transaction every two seconds. Each transaction has three write operations and two read operations. Each write operation triggers a change on a BO resulting in 5000 Bytes that have to be transferred to the secondary. We distinguish packet sizes of 100 Bytes (Workload A) and packet sizes of 1000 Bytes (Workload B). This means, the application message (15000 Bytes) has to be split into several network messages.

The figure shows the packet throughput for 100 Bytes (Figure 6 A) and 1000 Bytes (Figure 6 B) packet sizes. The time period for the measurement is 30 seconds. A packet size of 100 Bytes leads to a rather steady amount of 195 *pps* (packets per second). The variance using a 100 Byte packet size is roughly ± 10 *pps*. In contrast, 1000 Bytes leads to only 110 *pps* with a much higher variance. Having a small packet size where the application content is split in many network messages leads to a steady stream of small messages. The steady stream leads to a constant transfer rate of data. In contrast, large package sizes lead to a much more bursty behavior since message exchange is

---

[1] Semantics is also influenced by the parameters. For instance, using a lazy, update everywhere scheme in a WAN with high message is likely to lead to higher inconsistencies than in a LAN where updates are propagated faster.

concentrated to the time periods the application wants to send a message. Also, considerably less messages are sent in total.

Summarized, a small packet size produces an even, continuous and high packet stream which keeps network and server busy. However, peaks are not as high as with large message sizes. If the system is not able to manage the peaks created by large message sizes, small message sizes are preferable. The Byte throughput (bytes sent per time unit) behaves similarly.

6 Related Work

In the last decade, many different database replication strategies have been proposed [17,2,3,4]. Some of them integrate replica control directly into the database kernel while others use a middleware-based approach. That is, the replication logic is implemented in a layer between the client and the database replicas. The clients see only the middleware that provides a standard database interface such as JDBC. The middleware controls where reads are executed and makes sure that updates are performed at all replicas.

With regard to the Java platform, middleware solutions for group communication have been proposed such as Jgroup [18] which can serve as an additional reference in the case of realtime implementation. In J2EE components like session and entity beans keep the application state. The invocation of objects is intercepted by the server and transactions are handled by the transaction manager. Basically all application events are exposed to the application server, making a non-intrusive integration of an replication algorithm feasible. ADAPT [19] is a framework for application server replication providing an abstract view of the J2EE components in the server. It is based on the JBoss AS. Nevertheless, implementing a single algorithm using the framework is still likely to take months of developments.

Most industrial (such as WebSphere, Weblogic or JBoss) and many research solutions for application server replication [9,5,10,11] use the primary copy approach. In many cases, however, the AS replicas share a single database and/or the business objects are not replicated. The database cache is inactivated in many cases. The approaches are mainly designed for fault-tolerance. [12,13] provide replication of business objects but the represent algorithms are very specific.

7 Conclusions

This paper discusses replication alternatives for business objects at the application server layer, pointing out the differences of object replication at the application layer compared to traditional database replication.

We present a simulation framework that allows for a fast initial comparison of replication solutions before integrating them into an industrial system. A suitable parameter set for the simulation environment was collected. In our future work we are planning to perform a thorough comparison of algorithm alternatives and the impact of network configurations and workloads. We aim at developing an online advising system, making suggestions for reconfiguration, in order, e.g., to handle a change in workload.

References

1. Gray, J., Helland, P., O'Neil, P., Shasha, D.: The dangers of replication and a solution. In: SIGMOD Conf. (1996)
2. Pedone, F., Guerraoui, R., Schiper, A.: The database state machine approach. Distributed and Parallel Databases 14(1), 71–98 (2003)
3. Lin, Y., Kemme, B., Patiño-Martínez, M., Jiménez-Peris, R.: Middleware based data replication providing snapshot isolation. In: SIGMOD Conf. (2005)
4. Plattner, C., Alonso, G., Özsu, M.T.: Extending DBMSs with satellite databases. The VLDB Journal (2008)
5. Narasimhan, P., Moser, L.E., Melliar-Smith, P.M.: State synchronization and recovery for strongly consistent replicated CORBA objects. In: Int. Conf. on Dependable Systems and Networks, DSN (2001)
6. Killijian, M.O., Fabre, J.C.: Implementing a reflective fault-tolerant CORBA system. In: Int. Symp. on Reliable Distributed Systems, SRDS (2000)
7. Frølund, S., Guerraoui, R.: e-transactions: End-to-end reliability for three-tier architectures. IEEE Trans. Software Eng. 28(4), 378–395 (2002)
8. Othman, O., O'Ryan, C., Schmidt, D.C.: Strategies for CORBA middleware-based load balancing. In: IEEE(DS) Online (2001)
9. Felber, P., Narasimhan, P.: Reconciling replication and transactions for the end-to-end reliability of CORBA applications. In: Int. Symp. on Distributed Objects and Applications, DOA (2002)
10. Barga, R., Lomet, D., Weikum, G.: Recovery guarantees for general multi-tier applications. In: IEEE Int. Conf. on Data Engineering, ICDE (2002)
11. Wu, H., Kemme, B.: Fault-tolerance for stateful application servers in the presence of advanced transactions patterns. In: Int. Symp. on Reliable Distributed Systems, SRDS (2005)
12. Salas, J., Perez-Sorrosal, F., Patiño-Martínez, M., Jiménez-Peris, R.: WS-replication: a framework for highly available web services. In: Int. WWW Conf. (2006)
13. Perez-Sorrosal, F., Patiño-Martínez, M., Jiménez-Peris, R., Kemme, B.: Consistent and scalable cache replication for multi-tier J2EE applications. In: Int. Middleware Conf. (2007)
14. Wiesmann, M., Schiper, A., Pedone, F., Kemme, B., Alonso, G.: Database replication techniques: A three parameter classification. In: Int. Symp. on Reliable Distributed Systems, SRDS (2000)
15. Pedone, F., Wiesmann, M., Schiper, A., Kemme, B., Alonso, G.: Understanding replication in databases and distributed systems. In: IEEE Int. Conf. on Distributed Computing Systems, ICDCS (2000)
16. Sobeih, A., Hou, J.C., Kung, L.C., Li, N., Zhang, H., Chen, W.P., Tyan, H.Y., Lim, H.: J-sim: A simulation and emulation environment for wireless sensor networks. IEEE Wireless Communications 13(4) (2006)
17. Pacitti, E., Minet, P., Simon, E.: Fast algorithm for maintaining replica consistency in lazy master replicated databases. In: Int. Conf. on Very Large Data Bases, VLDB (1999)
18. Montresor, A.: Jgroup tutorial and programmer's manual (2000)
19. Babaoglu, Ö., Bartoli, A., Maverick, V., Patarin, S., Vuckovic, J., Wu, H.: A framework for prototyping J2EE replication algorithms. In: Int. Symp. on Distributed Objects and Applications, DOA (2004)

Applying Optimal Stopping for Optimizing Queries to External Semantic Web Resources

Albert Weichselbraun

Vienna University of Economics and Business Administration
Augasse 2-6, Vienna, Austria
albert.weichselbraun@wu-wien.ac.at
http://www.ai.wu-wien.ac.at/~aweichse

Abstract. The rapid increase in the amount of available information from various online sources poses new challenges for programs that endeavor to process these sources automatically and identify the most relevant material for a given application.

This paper introduces an approach for optimizing queries to Semantic Web resources based on ideas originally proposed by MacQueen for optimal stopping in business economics. Modeling applications as decision makers looking for optimal action/answer sets, facing search costs for acquiring information, test costs for checking these information, and receiving a reward depending on the usefulness of the proposed solution, yields strategies for optimizing queries to external services. An extensive evaluation compares these strategies to a conventional coverage based approach, based on real world response times taken from popular Web services.

1 Introduction

Semantic Web applications provide, integrate and process data from multiple data sources including third party providers. Combining information from different locations and services is one of the key benefits of semantic applications.

Current approaches usually limit their queries to a number of particularly useful and popular services as for instance Swoogle, GeoNames, or DBpedia. Research on automated web service discovery and matching [1] focuses on enhanced applications, capable of identifying and interfacing relevant resources in real time. Future implementations, therefore, could theoretically issue queries spawning vast collections of different data sources, providing even more enhanced and accurate information. Obviously, such query strategies - if applied by a large enough number of clients - impose a considerable load on the affected services, even if only small pieces of information are requested. The World Wide Web Consortium's (W3C) struggle against excessive document type definition (DTD) traffic provides a recent example of the potential impact a large number of clients achieves. Ted Guild pointed out (p.semanticlab.net/w3dtd) that the W3C receives up to 130 million requests per day from broken clients, fetching popular DTD's over and over again, leading to a sustained bandwidth consumption of approximately 350 Mbps.

J. Cordeiro et al. (Eds.): ICSOFT 2008, CCIS 47, pp. 105–118, 2009.

Service provider like Google restrict the number of queries processed on a per IP/user base to prevent excessive use of their Web services. From a client's perspective overloaded Web services lead to higher response times and therefore higher cost in terms of processing times and service outages.

Grass and Zilberstein [2] suggest applying value driven information gathering (VDIG) for considering the cost of information in query planning. VDIG focuses on the query selection problem in terms of the trade off between response time and the value of the retrieved information. In contrast approaches addressing only the coverage problem put their emphasis solely on maximizing precision and recall.

Optimizing value under scare resources is a classical problem from economics and highly related to decision theory. Applying these concepts to the information systems research domain yields important strategies for optimizing the acquisition of Web resources [3], addressing the trade-off between using resources sparingly and providing accurate and up-to-date information. In this research we apply the search test stop (STS) model to applications leveraging third party resources. The STS model considers the user's preferences between accuracy and processing time, maximizing the total utility in regard to these two measures. In contrast to the approach described by Grass and Zilberstein [2] the STS model adds support for a testing step, designed to obtain more information about the accuracy of the obtained results, aiding the decision algorithm in its decision whether to acquire additional information or act based on the current answer set. Similar to Ipeirotis et al. [4] the resulting query strategy might lead to less accurate results than a "brute force" approach, but nevertheless optimizes the balance between accuracy and costs. Therefore, the search test stop model addresses the trade-off between two of the major software engineering challenges outlined in ISO/IEC 9126-1: (i) *reliability* - the capability of the software product to maintain a level of accuracy according to measures specified in the software design process [5], and (ii) *efficiency* - requiring the software to provide an appropriate performance in terms of processing time and resource allocation, under stated conditions [6].

This paper's results are within the field of AI research facilitating techniques from decision theory to address problems of agent decision making [7].

The article is organized as follows. Section 2 presents known query limits and response times of some popular Web services. Section 3 provides the theoretical background for the search test stop model, and presents its extension to discrete probability functions. Afterwards the application of this method to applications utilizing external resources is outlined in Section 4 and an evaluation of this technique is presented in Section 5. This paper closes with an outlook and conclusions drawn in Section 6.

2   Performance and Scalability

The increased popularity of applications that rely on external data repositories calls for strategies for a responsible and efficient use of these resources.

Extensive queries to external resources increases their share of the program's execution time and may lead to longer response times, requiring its operators to impose limits on the service's use.

Even commercial providers like Google or Amazon restrict the number of accesses to their services. For instance, Google's Web API only allows 1000 requests a day, with

Table 1. Response times of some popular Web services

Service	Protocol	$\bar{t}_r$	$\tilde{t}_r$	t_r^{min}	t_r^{max}	$\sigma_{t_r}^2$
Amazon	REST	0.5	0.2	0.2	31.3	0.6
DBpedia	SPARQL	0.8	0.5	0.1	60.0	4.2
Del.icio.us	REST	0.6	0.4	0.1	24.3	0.5
GeoNames	REST	0.7	0.1	0.0	60.0	19.9
Google	Web	0.3	0.2	0.1	10.3	0.2
Swoogle	Web	4.1	1.6	0.2	60.0	98.4
Wikipedia	Web	0.5	0.2	0.1	60.0	1.3

exceptions for research projects. Workarounds like the use of Google's public Web interface may lead to blacklisting of the client's IP address[1]. Google's geo coding service imposes a limit of 15,000 queries per day and IP address. Amazon limits clients to 20 queries per second, but restrictions vary between the offered services and might change over time[2]. Other popular resources like GeoNames and Swoogle to our knowledge currently do not impose such limits.

A Web service timing application issuing five different queries to popular Web resources in 30 min intervals over a time period of five weeks yielded Table 2. The services' average response time ($\bar{t}_r$), the response time's median ($\tilde{t}_r$), its minimum and maximum values (t_r^{min}, t_r^{max}), and variance ($\sigma_{t_r}^2$) characterize its potential impact on an application's performance. Due to the timeout value of 60 seconds, specified in the timing application, all t_r^{max} values are equal or below 60. These response times vary, depending on the client's Internet connectivity and location, but adequate values can be easily obtained by probing the service's response times from the client's location.

Table 2 suggests that Google provides a fast and quite reliable service ($\sigma_{t_r}^2 = 0.2$) with only small variations in the response times. This result is not very surprising considering the global and highly reliable infrastructure Google employs.

Smaller information providers which cannot afford this kind of infrastructure in general provide good response times (due to fewer requests), but they are more sensitive to sudden peaks in the number of clients accessing their services as visualized in Figure 1. Table 2 reflects these spikes in terms of higher variances and t_r^{max} values.

Our experiments suggest (see Section 5) that especially clients querying services with high variances benefit from implementing the search test stop model.

Another strategy from the client's perspective is avoiding external resources at all. Many community projects like Wikipedia or GeoNames provide database dumps which might be used to install a local copy of the service. These dumps are usually rather large (a current Wikipedia dump including all pages, discussions, but without the edit history comprises approximately 7.8 GB[3]) and often outdated (Wikipedia dumps are sometimes even more than one month old, other services like GeoNames update their records very frequently).

---

[1] See p.semanticlab.net/gooso

[2] developer.amazonwebservices.com

[3] download.wikipedia.org; 2008-10-15

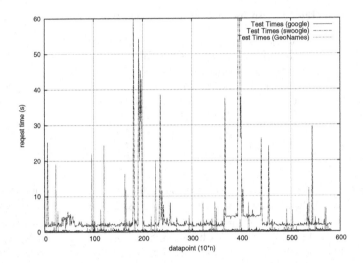

Fig. 1. Selected test times over the time, computed with a timeout of 60 seconds. Every data point accumulates five measurements.

The import of this data requires customized tools (like mwdumper[4]) or hacks and rarely processes without major hassles. In some cases the provided files do not contain all available data (GeoNames for instance does not publish the `relatedTo` information) so that querying the service cannot be avoided at all.

3   The Search Test Stop Model

This section outlines the basic principles of the search test stop (STS) model as found in decision theory. For a detailed description of the model please refer to MacQueen [8] and Hartmann [9].

MacQueen [8] describes the idea of the STS model as follows: A decision maker (a person or an agent) searches through a population of possible actions, sequentially discovering sets of actions (S_A), paying a certain cost each time a new set of actions is revealed (the search cost c_{s_i}). On the first encounter with a set of possible actions, the person obtains some preliminary information (x_0) about its utility (u), based on which he can

1. continue looking for another set of possible actions (paying search cost $c_{s_{i+1}}$),
2. test the retrieved set of actions, to obtain (x_1) - a better estimation of the actions value - paying the test cost (c_{t_i}) and based on this extended information continue with option 1 or finish the process with option 3, or
3. accept the current set of answers (and gain the utility u).

---

[4] www.mediawiki.org/wiki/MWDumper

The challenge is combining these three options so that the total outcome is optimized by keeping the search (c_{s_i}) and test (c_{t_i}) costs low ($\sum_{i=1}^{m} c_{s_i} + \sum_{i=1}^{n} c_{t_i}$) without jeopardizing the obtained utility u.

Introducing the transformation $r = E(u|x_0)$ yields the following description for a policy *without testing*:

$$v = vF(v) + \int_{v}^{+\infty} rf(r)dr - c_s \tag{1}$$

with the solution $v = v_0$. $F(r)$ represent the cumulative distribution function of the expected utility and $f(r)$ its probability mass function. The constant c_s refers to search cost and v (better v_0) to the utility obtained by the solution of this equation.

Extending Equation 1 to testing yields Equation 2:

$$v = vF(r_D) + \tag{2}$$
$$\int_{r_D}^{r_A} T(v,r)f(r)dr +$$
$$\int_{r_A}^{+\infty} rf(r)dr - c_s \quad \text{and}$$
$$T(v, r_D) = v \tag{3}$$
$$T(v, r_A) = r_A \tag{4}$$

$T(v,r)$ refers to the utility gained by testing, r_D to the value below which the discovered action set (S_A) will be dropped, and r_A to the minimal utility required for accepting S_A. A rational decision maker will only resort to testing, if the utility gained outweighs its costs and therefore the condition $T(v_0, v_0) > v_0$ holds which is the case in the interval $[r_D, r_A]$.

In the next two sections we will (i) describe the preconditions for applying this model to a real world use case, and (ii) present a solution for discrete data.

3.1 Preconditions

MacQueen [8] defines a number of preconditions required for the application of the STS model. Hartmann [9] eases some of these restrictions yielding the following set of requirements for the application of the model:

1. a common probability mass function $h(x_0, x_1, u)$ exists.
2. The expected value of u given a known realization x_0 ($z = E(U|x_0, y_0)$) exists and is finite.
3. $F(z|x_0)$ is stochastically increasing in x_0. For the concept of stochastically increasing variables please refer to [10, p75].

3.2 The Discrete Search Test Stop Model

This research deals with discrete service response time distributions and therefore applies the discrete STS methodology. Hartmann transferred MacQueen's approach to discrete models. The following section summarizes the most important points of his work [9].

Hartmann starts with a triple (x_0, x_1, u) of *discrete* probability variables, described by a common probability function $h(x_0, x_1, u)$. From h Hartmann derives

1. the conditional probability function $f(u|x_0, x_1)$ and the expected value $Z = E(u|x_0, x_1)$,
2. the probability function of r, $f(r|x_0)$ and $F(r|x_0)$,
3. the probability of x_0, $f(x_0)$ and $F(x_0)$.

Provided that the conditions described in Section 3.1 are fulfilled only five possible optimal policies are possible - (i) always test, (ii) never test, (iii) test if $u > u_t$, (iv) if $u < u_t$, or (v) if $u_t < u < u_t'$.

The expected utility equals to

1. $E(u|x_0)$ for accepting without testing,
2. $T(r, v)$ with testing, and
3. v_0 if the action is dropped and a new set (S_A) is selected according to the optimal policy.

4   Method

This section focuses on the application of the STS model to Web services. At first we describe heuristics for estimating cost functions (c_s, c_t), and the common probability mass function $h(x_0, x_1, u)$ Afterwards the process of applying search test stop to tagging applications is elaborated.

4.1   Cost Functions

In the conventional STS model costs refer to the investment in terms of time and money for gathering information. By applying this idea to software, costs comprise all expenses in terms of CPU-time, bandwidth and storage cost necessary to search for or test certain answers.

Large scale Semantic Web projects, like the IDIOM media watch on climate change [11], process hundred thousands of pages a week. Querying GeoNames for geo-tagging such numbers of documents would add *days* of processing time to the IDIOM architecture.

This research focuses solely on costs in terms of response time, because they are the limiting factor in our current research projects. Other applications might require extending this approach to consider additional cost factors like CPU-time, bandwidth, etc.

4.2   Utility Distributions

Applying the STS model to economic problems yields cash deposits and payments. Transferring this idea to information science is a little bit more subtle, because the utility is highly dependent on the application and its user's preferences. Even within one domain the notion of an answer set's (S_A) value might not be clear. For instance in a geo context the "correct" answer for a certain problem may be a particular mountain in Austria, but the geo-tagger might not identify the mountain but the surrounding region or at least the state in which it is located (compare Figure 2).

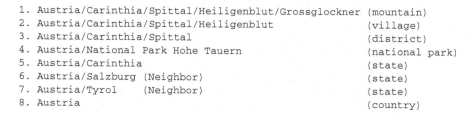

```
1. Austria/Carinthia/Spittal/Heiligenblut/Grossglockner (mountain)
2. Austria/Carinthia/Spittal/Heiligenblut              (village)
3. Austria/Carinthia/Spittal                           (district)
4. Austria/National Park Hohe Tauern                   (national park)
5. Austria/Carinthia                                   (state)
6. Austria/Salzburg (Neighbor)                         (state)
7. Austria/Tyrol    (Neighbor)                         (state)
8. Austria                                             (country)
```

Fig. 2. Ranking of "correct" results for geo-tagging an article covering the "Grossglockner"

Assigning concrete utility values to these alternatives is not possible without detailed information regarding the application and user preferences. Approaches for evaluating the set's value might therefore vary from binary methods (full score for correct answers; no points for incomplete/incorrect answers) to complex ontology based approaches, evaluating the grade of correctness and severe of deviations.

4.3 Application

This work has been motivated by performance issues in a geo-tagging application facilitating resources from GeoNames and WordNet for improving tagging accuracy. Based on the experience garnered during the evaluation of STS models, this section will present a heuristic for determining the cost functions (c_s, c_t) and the common probability mass function $h(x_0, x_1, u)$.

Figure 3 visualizes the application of the search test stop model to Web services. Searching yields an answer set $S_a = \{a_1, \ldots, a_n\}$ and the indicator x_0 at a prices of c_s. Based on x_0 the search test stop description logic decides on whether to (i) accept the current answer set, (ii) drop the answer set and continue searching, or (iii) query another set of resources to retrieve the refined indicator x_1 paying the test cost c_t. Based on x_1 the answer set is dropped or finally accepted.

Cost Functions. Searching leads to external queries and therefore costs. Measuring a service's performance over a certain time period allows estimating the average response time and variance.

STS fits best for situations, where the query cost c_s is in the same order as the average utility retrieved ($O(c_s) = O(\overline{u})$). In settings with $O(c_s) \ll O(\overline{u})$ the search costs have no significant impact on the utility and if $O(c_s) \gg O(\overline{u})$ no searching will take place at all (because the involved costs are much higher than the possible benefit).

In real world situations the translation from search times to costs is highly user dependent. To simplify the comparison of the results, this research applies a linear translation function $c_s = \lambda \cdot t_s$ with $\lambda = const = 1/\tilde{t}_s$ yielding costs of $O(c_s) = 1$. Selecting

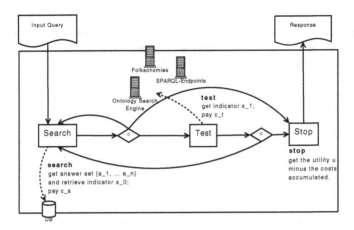

Fig. 3. Applying the search test stop model to Web resources

the median of the response times $\tilde{t}_s$ and specifying a timeout value of 60 seconds for any query operation reduces the influence of service outages on the simulation results.

The performance of the search test stop algorithm is highly dependent on accurate estimations of the query cost, because all decisions are solely based on the common probability mass function and the cost functions. Future research will compute query cost based on use case specific cost functions as demonstrated by Strunk et al. [12], Verma et al. [13], and Yeo and Buyya [14] and evaluate the results yielded by these different approaches.

Utility Distribution. The discrete common probability mass function h is composed of three components: The probability mass function of (i) the utility u, (ii) the random variable x_0 providing an estimate of the utility and, (iii) the random variable x_1 containing a refined estimate of the answer's utility.

In general a utility function assuming linearly independent utility values might look like Equation 5.

$$u = \sum_{a_i \in S_A} \lambda(a_i) f_{eval}(a_i) \tag{5}$$

The utility equals to the sum of the utility gained by each answer set S_A, which is evaluated using an evaluation function f_{eval}, and weighted with a factor $\lambda(a_i)$. To simplify the computation of the utility we consider only correct answers as useful (Equation 6) and apply the same weight ($\lambda(a_i) = const = 1$) to all answers.

$$f_{eval}(a_i) = \begin{cases} 0 & \text{if } a_i \text{ incorrect;} \\ 1 & \text{if } a_i \text{ correct.} \end{cases} \tag{6}$$

Geo-tagging identifies geographic entities based on a knowledge base as for instance a gazetteer or a trained artificial intelligence algorithm.

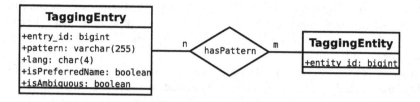

Fig. 4. Database schema of a simple tagger

After searching the number of identified entries ($|S_a| = x_0$) provides a good estimation of the expected value of the answers utility. Applying a focus algorithm (e.g. [15]) yields a refined evaluation of the entity set ($|S'_a| = x_1$) resolving geo ambiguities. S'_a might still contain incorrect answers due to errors in the geo disambiguation or due to ambiguous terms not resolved by the focus algorithm (e.g. turkey/bird versus Turkey/country). Based on the probabilities of a particular answer $a_i \in S_a/a'_i \in S'_a$ of being incorrect $P_{incorr}(a_i)/P_{incorr}(a'_i)$ the expected value u for a given combination of x_0, x_1 is determined. Evaluating historical error rates yields estimations for $P_{incorr}(a_i)$ and $P_{incorr}(a'_i)$.

If no historical data is available heuristics based on the number of ambiguous geo-entries are useful for providing an educated guess of the probabilities.

A tagger recognizes patterns based on a pattern database table. The relation has Pattern translates these patterns to TaggingEntities as for instance spatial locations, persons, and organizations. Figure 4 visualizes a possible database layout for such a tagger.

The hasPattern table often does not provide a unique mapping between patterns and entities - names as for instance Vienna may refer to multiple entities (Vienna/Austria versus Vienna/Virgina/US). On the other side many entities have multiple patterns associated with them (e.g. Wien, Vienna, Vienne, Bech, etc.). Based on the database schema above, $P_{incorr}(a_i)$ for such a tagger is estimated using the following heuristic:

$$n_{Entities} = |TaggingEntity| \tag{7}$$

$$n_{Mappings} = |hasPattern| \tag{8}$$

$$n_{ambiguous} = |\sigma_{[isAmbiguous='true']}(\tag{9}$$
$$TaggingEntry * hasPattern)|$$

$$P_{incorr} = 1 - \frac{n_{Entries}}{n_{Mappings} + n_{ambiguous}} \tag{10}$$

Extending the database schema visualized in Figure 4 to non geo entries using WordNet and applying Equations 7-10 yields $P_{incorr}(a'_i)$.

5   Evaluation

For evaluating the STS model's efficiency in real world applications a simulation framework, supporting (i) a solely coverage based decision logic and the search test stop model, (ii) artificial (normal distribution) and measured (compare Section 2) distributions of network response times, and (iii) common probability mass functions $h(x_0, x_1, u)$ composed from user defined $P_{incorr}(a_i)$ and $P_{incorr}(a_i')$ settings have been programmed.

To prevent the coverage based decision logic from delivering large amounts of low quality answers, the simulation controller only accepts answers with an expected utility above a certain threshold (u_{min}). In contrast the search test stop algorithm computes $u_{min} = r_D$ on the fly, based on the current responsiveness of the external service and the user's preferences.

5.1 Performance

Comparing the two approaches at different minimum quality levels (u_{min}), and service response time distributions approximated by a normal distribution $N(\bar{t}, \sigma_t^2)$ yields Table 2. The common probability mass functions has been composed with $P_{incorr}(a_i) = 0.3$, $P_{incorr}(a_i') = 0.1$. The parameters for the normal distribution are $c_s = N(2, 1.9)$ for high search costs, $c_s = N(1, 0.9)$ for medium search costs, and $c_s = N(0.5, 0.4)$ for low search costs.

Table 2. Tagging performance

Search		Quality ($\bar{u}$)		Quantity ($\frac{\Delta u}{\Delta t}$)	
Cost (c_s)	u_{min}	STS	Conv	STS	Conv
low	2	6.62	5.58	3.47	7.79
low	4	6.64	6.13	3.56	6.93
low	6	6.69	6.55	3.57	5.95
low	8	6.66	6.39	3.55	2.75
medium	2	4.99	4.84	1.88	3.22
medium	4	5.02	5.15	1.92	2.76
medium	6	5.01	5.32	1.89	2.27
medium	8	5.00	3.86	1.87	0.79
high	2	2.81	3.20	0.78	1.05
high	4	2.75	3.25	0.76	0.88
high	6	2.84	2.81	0.80	0.59
high	8	2.81	-0.91	0.76	-0.09

Table 2 evaluates the two strategies according to two criteria: (i) *answer quality* $\bar{u}$, the average utility of a set (S_A) retrieved by the strategy, and (ii) *answer quantity* $\frac{\Delta u}{\Delta t}$, the rate at which the number of correct answers (and therefore the total utility (u)) grows.

High $\bar{u}$ values correspond to accepting only high quality results, with a lot of correct answers, and dropping low quality answer sets (at the cost of a lower quantity).

The conventional coverage based approach (Conv) delivers the highest quantity for small u_{min} values because virtually all answers are accepted and contribute to the total utility. This greedy approach comes at the cost of a lower answer quality and therefore low average utility $\bar{u}$ per answer. Increasing u_{min} yields a better answer quality, but lower quantity values. At high search costs this strategy's performance is particularly unsatisfactory, because it doesn't consider the costs of the search operation.

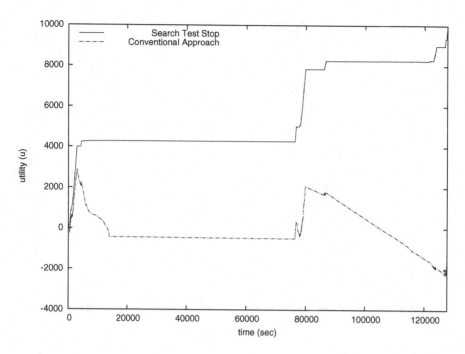

Fig. 5. Search test stop versus conventional decision logic for Swoogle ($\tilde{t}$=1.6; $\sigma^2_{t_r} > 10000$)

In contrast to the conventional approach STS maximizes answer quality and quantity based on the *current* search cost adjusting queries to the responsiveness of the service and the user's preferences. These preferences formalize the trade-off between quality and quantity by specifying a transformation function between search cost and search times.

STS therefore optimizes the agent's behavior in terms of user utility. This does *not* mean that STS minimizes resource usage. Instead STS dynamically adjusts the resource utilization based on the cost of searching (c_s) and testing (c_t), providing the user with optimal results in terms of accuracy *and* response times.

Enforcing a minimal utility u_{min} boosts the average utility $\overline{u}$ of the non STS service, but at the cost of a higher resource utilization, independent from the server's load (leading to extremely high response times during high load conditions). Static limits also do not consider additional queries at idle servers, leading to lower utilities under low load conditions. In contrast to the conventional approach STS (i) utilizes dormant resources of idle servers, and (ii) spares resources of busy servers, maximizing utility according to the user's preferences.

5.2 Web Services

In this section we will simulate the effect of STS on the performance of real world Web services, using search costs as measured during the Web service timing (compare Section 2).

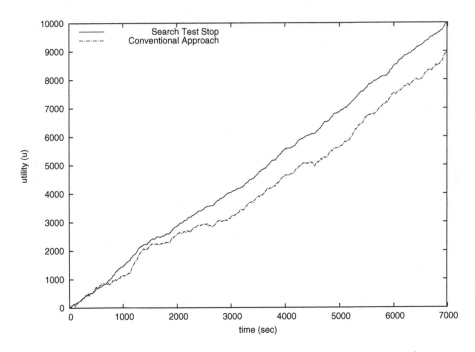

Fig. 6. Search test stop versus the conventional decision logic for Google ($\tilde{t}=0.2$; $\sigma^2_{t_r}=0.2$)

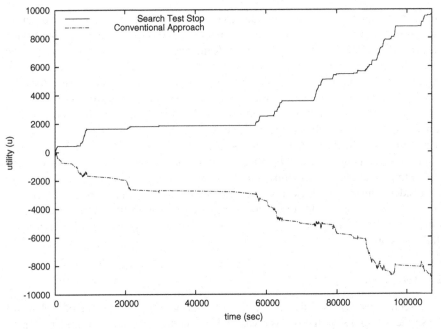

Fig. 7. Search test stop versus the conventional decision logic for GeoNames ($\tilde{t}=0.1$; $\sigma^2_{t_r}=771.4$)

The simulation facilitates the cost and common probability mass functions from Section 5. The figures 5-7 compare the tagger's performance when providing tagged documents corresponding to a utility score of 10,000 based on three different Web services (Swoogle, Google, GeoNames) with a minimum utility (u_{min}) of four.

In all three use cases STS performs well, because the search times are adjusted according to the service's responsiveness. GeoNames and Swoogle experience the highest performance boost, due to high variances in the search cost, leading to negative utility for the conventional query strategy.

Using Google as external resource yields the fastest processing time. The algorithm is able to provide documents with the required quality level in around 8,300 seconds in contrast to more than 107,000 seconds for GeoNames and more than 129,000 seconds for Swoogle.

Services with low variances ($\sigma_{t_r}^2$) in their response times as for instance Google, del.icio.us and Wikipedia benefit least from the application of the STS model, because static strategies perform reasonable well under these conditions.

6   Outlook and Conclusions

This work presents an approach for optimizing access to third party remote resources. Optimizing the clients resource access strategy yields higher query performance and spares remote resources by preventing unnecessary queries.

The main contributions of this paper are (i) applying the search test stop model to value driven information gathering, extending its usefulness to domains where one or more testings steps allow refining the estimated utility of the answer set; (ii) demonstrating the use of this approach to semantic tagging, and (iii) evaluating how the search test stop model performs in comparison to a solely value based approach.

The experiments show that search test stop and value driven information gathering perform especially well in domains with highly variable search cost.

In this work we only use one level testing, nevertheless, as Hartmann has shown [9] extending STS to n-levels of testing is a straight forward task. Future research will transfer these techniques and results to more complex use cases integrating multiple data sources as for instance semi automatic ontology extension [16]. The development of utility functions considering partially correct answers and user preferences will allow a more fine grained control over the process's performance yielding highly accurate querying strategies and therefore better results.

Acknowledgements. The author wishes to thank Prof. Wolfgang Janko for his valuable feedback and suggestions. The project results have been developed in the IDIOM (Information Diffusion across Interactive Online Media; www.idiom.at) project funded by the Austrian Ministry of Transport, Innovation & Technology (BMVIT) and the Austrian Research Promotion Agency (FFG).

References

1. Gupta, C., Bhowmik, R., Head, M.R., Govindaraju, M., Meng, W.: Improving performance of web services query matchmaking with automated knowledge acquisition. In: Web Intelligence, pp. 559–563. IEEE Computer Society, Los Alamitos (2007)

2. Grass, J., Zilberstein, S.: A value-driven system for autonomous information gathering. Journal of Intelligent Information Systems 14(23), 5–27 (2000)
3. Kukulenz, D., Ntoulas, A.: Answering bounded continuous search queries in the world wide web. In: WWW 2007: Proceedings of the 16th international conference on World Wide Web, pp. 551–560. ACM, New York (2007)
4. Ipeirotis, P.G., Agichtein, E., Jain, P., Gravano, L.: Towards a query optimizer for text-centric tasks. ACM Trans. Database Syst. 32(4), 21 (2007)
5. Software Engineering Standard Committe of the IEEE Computer Society: IEEE std 830-1999: IEEE recommended practice for software requirements specifications (1998)
6. International Standards Organization JTC 1/SC 7: ISO/IEC 9126-1, 2001. software engineering – product quality – part 1: Quality model (2001)
7. Horvitz, E.J., Breese, J.S., Henrion, M.: Decision theory in expert systems and artificial intelligence. International Journal of Approximate Reasoning 2, 247–302 (1988)
8. MacQueen, J.: Optimal policies for a class of search and evaluation problems. Management Science 10(4), 746–759 (1964)
9. Hartmann, J.: Wirtschaftliche Alternativensuche mit Informationsbeschaffung unter Unsicherheit. PhD thesis, Universität Fridericiana Karlsruhe (1985)
10. Lehmann, E.L., Romano, J.P.: Testing Statistical Hypotheses, 3rd edn. Springer, New York (2005)
11. Scharl, A., Weichselbraun, A., Liu, W.: Tracking and modelling information diffusion across interactive online media. International Journal of Metadata, Semantics and Ontologies 2(2), 136–145 (2007)
12. Strunk, J.D., et al.: Using utility to provision storage systems. In: FAST 2008: Proceedings of the 6th USENIX Conference on File and Storage Technologies, pp. 1–16. USENIX Association, Berkeley (2008)
13. Verma, A., Jain, R., Ghosal, S.: A utility-based unified disk scheduling framework for shared mixed-media services. Trans. Storage 3(4), 1–30 (2008)
14. Yeo, C.S., Buyya, R.: Pricing for utility-driven resource management and allocation in clusters. International Journal of High Performance Computing Applications 21(4), 405–418 (2007)
15. Amitay, E., Har'El, N., Sivan, R., Soffer, A.: Web-a-where: geotagging web content. In: SIGIR 2004: Proceedings of the 27th annual international ACM SIGIR conference on Research and development in information retrieval, pp. 273–280. ACM, New York (2004)
16. Liu, W., Weichselbraun, A., Scharl, A., Chang, E.: Semi-automatic ontology extension using spreading activation. Journal of Universal Knowledge Management 1, 50–58 (2005), http://www.jukm.org/jukm_0_1/semi_automatic_ontology_extension

An Efficient Pipelined Parallel Join Algorithm on Heterogeneous Distributed Architectures

Mohamad Al Hajj Hassan and Mostafa Bamha

LIFO, University of Orléans, BP 6759, 45067 Orléans cedex 2, France
alhassan@univ-orleans.fr, bamha@univ-orleans.fr

Abstract. Pipelined parallelism was largely studied and successfully implemented, on shared nothing machines, in several join algorithms in the presence of ideal conditions of load balancing between processors and in the absence of data skew. The aim of pipelining is to allow flexible resource allocation while avoiding unnecessary disk input/output for intermediate join results in the treatment of multi-join queries.

The main drawback of pipelining in existing algorithms is that communication and load balancing remain limited to the use of static approaches (generated during query optimization phase) based on hashing to redistribute data over the network and therefore cannot solve data skew problem and load imbalance between processors on heterogeneous multi-processor architectures where the load of each processor may vary in a dynamic and unpredictable way.

In this paper, we present a pipelined parallel algorithm for multi-join queries allowing to solve the problem of data skew while guaranteeing perfect balancing properties, on heterogeneous multi-processor Shared Nothing architectures. The performance of this algorithm is analyzed using the scalable portable BSP (Bulk Synchronous Parallel) cost model.

1 Introduction

The appeal of parallel processing becomes very strong in applications which require ever higher performance and particularly in applications such as : data-warehousing, decision support, OLAP[1] and more generally DBMS[2] [12,7]. However parallelism can only maintain acceptable performance through efficient algorithms realizing complex queries on dynamic, irregular and distributed data. Such algorithms must be designed to fully exploit the processing power of multi-processor machines and the ability to evenly divide load among processors while minimizing local computation and communication costs inherent to multi-processor machines.

Research has shown that the join operation is parallelizable with near-linear speed-up on Shared Nothing machines only under ideal balancing conditions. Data skew can have a disastrous effect on performance[14,11,8,3,4] due to the high costs of communications and synchronizations in this architecture.

Many algorithms have been proposed to handle data skew for a simple join operation, but little is known for the case of complex queries leading to multi-joins [13,8,11].

---

[1] OLAP : On-Line Analytical Processing.
[2] PDBMS : Parallel Database Management Systems.

J. Cordeiro et al. (Eds.): ICSOFT 2008, CCIS 47, pp. 119–133, 2009.
© Springer-Verlag Berlin Heidelberg 2009

In particular, the performance of PDBMS has generally been estimated on queries involving one or two join operations only [19]. However the problem of data skew is more acute with multi-joins because the imbalance of intermediate results is unknown during static query optimization [13]. Such algorithms are not efficient for many reasons:

- the presented algorithms cannot be scalable (and thus cannot guarantee linear-speedup) because their routing decisions are generally performed by a coordinator processor while the other processors are idle,
- they cannot solve load imbalance problem as they base their routing decisions on incomplete or statistical information,
- they cannot solve data skew problem because data redistribution is generally based on static hashing data into buckets and data hashing is known to be inefficient in the presence of high frequencies [8].

On the contrary, the join algorithms we presented in [4,3,1], use a total data-distribution information in the form of histograms. The parallel cost model we apply allows us to guarantee that histogram management has a negligible cost when compared to the efficiency gains it provides to reduce the communication cost and to avoid load imbalance between processors.

In homogeneous multi-processor architectures, these algorithms are insensitive to the problem of data skew and guarantee perfect load balancing between processors during all the stages of join computation because data redistribution is carried out jointly by all processors (and not by a coordinator processor). Each processor deals with the redistribution of the data associated to a subset of the join attribute values, not necessarily its "own" values. However the performance of these algorithms degrades on heterogeneous multi-processor architectures where the load of each processor may vary in a dynamic and unpredictable way.

To this end we introduced in [10] a parallel join algorithm called *DFA-Join* (Dynamic Frequency Adaptive parallel join algorithm) to handle join queries on heterogeneous Shared Nothing architectures allowing to solve the problem of data skew while guaranteeing perfect balancing properties.

In this paper, we present a pipelined of version of *DFA-Join* join algorithm called *PDFA-Join* (Pipelined Dynamic frequency Adaptive join algorithm). The aim of pipelining in *PDFA-Join* is to offer flexible resource allocation and to avoid unnecessary disk input/output for intermediate join result in multi-join queries. We show that *PDFA-Join* algorithm can be applied efficiently in various parallel execution strategies making it possible to exploit not only intra-operator parallelism but also inter-operator parallelism. These algorithms are used in the objective to avoid the effect of load imbalance due to data skew, and to reduce the communication costs due to the redistribution of the intermediate results which can lead to a significant degradation of the performance.

2   Limitations of Parallel Execution Strategies in Multi-join Queries

Parallel execution of multi-join queries depends on the execution plan of simple joins that compose it. The main difference between these strategies lies in the manner of allocating the simple joins to different processors and in the choice of an appropriate degree of parallelism (i.e. the number of processors) used to compute each simple join.

Several strategies were proposed to evaluate multi-join queries [12,19]. They generally depend on the parallel query execution plan. In these strategies intra-operator, inter-operator and pipelined parallelisms can be used. These strategies are divided into four principal categories presented thereafter. We will give for each one its advantages and limitations.

2.1 Sequential Parallel Execution

Sequential parallel execution is the simplest strategy to evaluate, in parallel, a multi-join query. It does not induce inter-operator parallelism. Simple joins are evaluated one after the other in a parallel way. Thus, at a given moment, one and only one simple join is computed in parallel by all the available processors.

This strategy is very restrictive and does not provide efficient resource allocation owing to the fact that a simple join cannot be started until all its operands are entirely available, and whenever a join operation is executed on a subset of processors, all the other processors remain idle until the next join operation. Moreover this strategy induces unnecessary disk Input/Output because intermediate results are written to disk and not immediately used for the next operations.

The execution time of each join is then the execution time of the slowest processor. Figure 1 illustrates an example of a sequential parallel execution. To reach acceptable performance, join algorithms used in this strategy should reduce the load imbalance between all the processors and the number of idle processors must be as small as possible.

2.2 Parallel Synchronous Execution

Parallel synchronous execution uses in addition to intra-operator parallelism, inter-operator parallelism [6]. In this strategy several simple join operations can be computed simultaneously on disjoint sets of processors. Figure 2 gives an example of a parallel synchronous execution: Processors 1 to 4 compute join 1 while processors 5 to 10 compute join 2. Once these joins finish, all the processors are used to compute join 3 and then join 4.

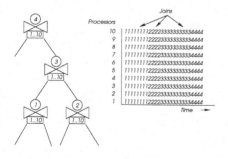

Fig. 1. Sequential parallel execution

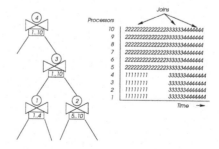

Fig. 2. Parallel synchronous execution

The parallel execution time of an operator depends on the degree of parallelism. The execution time decreases by increasing the number of processors until the arrival at a point of saturation (called optimal degree of parallelism) from which increasing the number of processors, increases the parallel execution time [15,6]. The main difficulty in this strategy lies in the manner of allocating the simple joins to the available processors and in the choice of an appropriate degree of parallelism to be used for each join.

In this strategy, the objective of such allocation is to reduce the latency so that the global execution time of all operators should be of the same order. This also applies to the global execution time of each operator in the same group of processors where the local computation within each group must be balanced.

This Strategy combines only intra- and inter operator parallelism in the execution of multi-join queries and does not introduce pipelined parallelism and large number of processors may remain idle if aren't used in inter-operator parallelism. This constitutes the main limitations of this strategy for flexible resource allocation in addition to unnecessary disk/input operation for intermediate join result.

2.3   Segmented Right-Deep Execution

Contrary to a parallel synchronous strategy, a *Segmented Right-Deep execution* [5,12] employs, in addition to intra-operator parallelism, pipelined inter-operator parallelism which is used in the evaluation of the right-branches of the query tree.

An example of a segmented right-deep execution is illustrated in figure 3 where all the available processors initially compute join 1 and disjoint sub-sets of processors are used to compute simultaneously joins 2, 3 and 4 using pipelined parallelism. Note that pipelined parallelism cannot start until the creation of the hash table of the join result of operation 1.

Segmented right-deep execution offers more flexible resource allocation than parallel synchronous execution strategy : many joins can be computed on disjoint sets of processors to prepare hash tables for pipelined joins. Its main limitation remains in the fact that pipelined parallelism cannot be started until all the hash tables are computed. Moreover no load balancing between processors can be performed whenever pipelined parallelism begins.

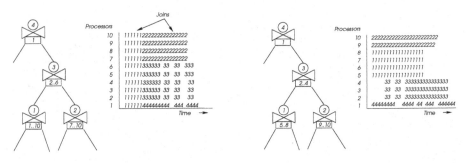

Fig. 3. Segmented right-deep execution   **Fig. 4.** Full parallel execution

2.4   Full Parallel Execution

Full Parallel execution [18,19,12] uses inter-operator parallelism and pipelined inter-operator parallelism in addition to intra-operator parallelism. In this strategy, all the simple joins, associated to the multi-join query, are computed simultaneously in parallel using disjoint sets of processors. Inter-operator parallelism and pipelined inter-operator parallelism are exploited according to the type of the query tree.

The effectiveness of such strategy depends on the quality of the execution plans generated during the query optimization phase and on the ability to evenly divide load between processors in the presence of skewed data.

All existing algorithms using this strategy are based on static hashing to redistribute data over the network which makes them very sensitive to data skew. Moreover pipelined parallelism cannot start until the creation of hash tables of build relations. We recall that all join algorithms used in these strategies require data redistribution of all intermediate join results (and not only tuples participating to the join result) which may induce a high cost of communication. In addition no load balancing between processors can be performed when pipelined parallelism begins, this can lead to a significant degradation of performance.

In the following section we will present *PDFA-Join* (Pipelined Dynamic Frequency Adaptive Join): a new join algorithm which can be used in different execution strategies allowing to exploit not only intra-operator but also inter-operator and pipelined parallelism. This algorithms is proved to induce a minimal cost for communication (only relevant tuples are redistributed over the network), while guaranteeing perfect load balancing properties in a heterogeneous multi-processor machine even for highly skewed data.

3   Parallelism in Multi-join Queries Using PDFA-Join Algorithm

Pipelining was largely studied and successfully implemented in many classical join algorithms, on Shared Nothing (SN) multi-processor machine in the presence of ideal conditions of load balancing and in the absence of data skew [12]. Nevertheless, these algorithms are generally based on static hash join techniques and are thus very sensitive to AVS and JPS.

The pipelined algorithm we introduced in [2] solves this problem and guarantees perfect load balancing on homogeneous SN machines. However its performance degrades on heterogeneous multi-processor architectures. where the load of each processor may vary in a dynamic and unpredictable way.

In this paper, we propose to adapt *DFA-Join* to pipelined multi-join queries to solve the problem of data skew and load imbalance between processors on heterogeneous multi-processors architectures.

DFA-Join is based on a combination of two steps :

- a static step where data buckets are assigned to each processor according to its actual capacity,
- then a dynamic step executed throughout the join computation phase to balance load between processors. When a processor finishes join processing of its assigned buckets it asks a local head node for untreated buckets of another processor.

This combination of static and dynamic steps allows us to reduce the join processing time because in parallel systems the total executing time is the time taken by the slowest processor to finish its tasks.

To ensure the extensibility of the algorithm, processors are partitioned into disjoint sets. Each set has a designed local head node. Load is first balanced inside each set of processors and whenever a set of processors finishes its assigned tasks, it asks a head node of another set of processors for additional tasks.

3.1 Detailed Algorithm

In this section, we present a parallel pipelined execution strategy for the multi-join query, $Q = (R \bowtie_{a_1} S) \bowtie_{b_1} (U \bowtie_{a_2} V)$, given in figure 5 (this strategy can be easily generalized to any bushy multi-join query) where R, S, U and V are source relations and a_1, a_2 and b_1 are join attributes.

We will give in detail the execution steps to evaluate the join query $Q_1 = R \bowtie_{a_1} S$ (the same technique is used to evaluate $Q_2 = U \bowtie_{a_2} V$). We first assume that each

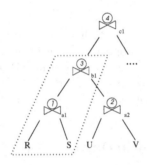

Fig. 5. Parallel execution of a multi-join query using PDFA_join algorithm

relation $T \in \{R, S, U, V\}$ is horizontally fragmented among p processors and:

- T_i denotes the fragment (sub-relation) of relation T placed on processor i,
- $Hist^x(T)$ denotes the histogram[3] of relation T with respect to the join attribute x, i.e. a list of pairs (v, n_v) where $n_v \neq 0$ is the number of tuples of relation T having the value v for the join attribute x. The histogram is often much smaller and never larger than the relation it describes,
- $Hist^x(T_i)$ denotes the histogram of fragment T_i placed on processor i,
- $Hist_i^x(T)$ is processor i's fragment of the global histogram of relation T,
- $Hist^x(T)(v)$ is the frequency (n_v) of value v in relation T,
- $Hist^x(T_i)(v)$ is the frequency of value v in sub-relation T_i,
- $\|T\|$ denotes the number of tuples of relation T, and
- $|T|$ denotes the size (expressed in bytes or number of pages) of relation T.

---

[3] Histograms are implemented as balanced trees (B-tree): a data structure that maintains an ordered set of data to allow efficient search and insert operations.

Algorithm 1. Parallel PDFA-Join computation steps to evaluate the join of R and S on attribute a_1 and preparing the next join on attribute b_1

In Parallel   (on each processor) $i \in [1, p]$ **do**

1 ▶ Create local histograms $Hist^{a_1}(R_i)$ of relation R_i and, on the fly, hash the tuples of relation R_i into different buckets according to the values of join attribute a_1,

  ▷ Create local histograms $Hist^{a_1}(S_i)$ of relation S_i and, on the fly, hash the tuples of relation S_i into different buckets according to the values of join attribute a_1,

2 ▶ Create global histogram fragment's, $Hist_i^{a_1}(R)$ of relation R on each processor i,

  ▷ Create global histogram fragment's, $Hist_i^{a_1}(S)$ of relation S on each processor i,

  ▷ Merge $Hist_i^{a_1}(R)$ and $Hist_i^{a_1}(S)$ to create join histogram, $Hist_i^{a_1}(R \bowtie S)$ on each processor i,

3 ▶ Create communication templates for only tuples participating to join result,

  ▷ Filter generated buckets to create tasks to execute on each processor according to its capacity,

  ▷ Create local histograms $Hist^{b_1}(R_i \bowtie S_i)$ of join result (of the buckets associated to processor i)
  on attribute b_1 of the next join using histograms and communication templates
  (See Algo. 2.),

4 ▶ Exchange data stored on each bucket according to communication templates,

5 ▶ Execute join tasks (of each bucket) on each processor, and store the join result on local disk.

  Loop until no task to execute
    ▷ Ask a local head node for jobs from an overloaded processor,
    ▷ Steal a job from a designated processor and execute it,
    ▷ Store the join result on local disk.
  End Loop
EndPar

Our algorithm (*Algorithm 1*) can be divided into the following five phases. We give for each phase an upper bound of the execution time using BSP (Bulk Synchronous Parallel) cost model [16,17]. Notation $O(...)$ only hides small constant factors : they depend only on the implementation, but neither on data nor on the BSP machine parameters.

Phase 1. Creating local histograms:
In this phase, we create in parallel, on each processor i, the local histogram $Hist^{a_1}(R_i)$ (resp. $Hist^{a_1}(S_i)$) $(i = 1, \ldots, p)$ of block R_i (resp. S_i) by a linear traversal of R_i (resp. S_i) in time $\max_{i=1,\ldots,p}(c_{r/w}^i * |R_i|)$ (resp. $\max_{i=1,\ldots,p}(c_{r/w}^i * |S_i|)$) where $c_{r/w}^i$ is the cost to read/write a page of data from disk on processor i.

While creating the histograms, tuples of R_i (resp. S_i) are partitioned on the fly into N buckets using a hash function in order to facilitate the redistribution phase. The cost of this phase is:

$$Time_{phase1} = O\left(\max_{i=1,\ldots,p} c_{r/w}^i * (|R_i| + |S_i|) \right).$$

Phase 2. Computing the histogram of $R \bowtie S$:
In this phase, we compute $Hist_i^{a_1}(R \bowtie S)$ on each processor i. This helps in specifying the values of the join attribute that participate in the join result, so only tuples of R and S related to these values are redistributed in a further phase which allows us to minimize the communication cost. The histogram of $R \bowtie S$ is simply the intersection of $Hist^{a_1}(R)$ and $Hist^{a_1}(S)$, so we must first compute the global histograms $Hist_i^{a_1}(R)$ and $Hist_i^{a_1}(S)$ by redistributing the tuples of the local histograms using a hash function that distributes the values of the join attribute in a manner that respects the processing capacity of each processor. The cost of this step is:

$$Time_{phase2.a} =$$
$$O\Big(\min \big(\max_{i=1,\dots,p} \omega_i * p * (g * |Hist^{a_1}(R)| + \gamma_i * ||Hist^{a_1}(R)||),$$
$$\max_{i=1,\dots,p} \omega_i * (g * |R| + \gamma_i * ||R||) \big) + \min \big(\max_{i=1,\dots,p} \omega_i * p * (g * |Hist^{a_1}(S)| +$$
$$\gamma_i * ||Hist^{a_1}(S)||), \max_{i=1,\dots,p} \omega_i * (g * |S| + \gamma_i * ||S||) \big) + l \Big).$$

where ω_i is the fraction of the total volume of data assigned to processor i such that: $\omega_i = (\frac{1}{\gamma_i})/(\sum_{j=1}^{p} \frac{1}{\gamma_j})$, γ_i is the execution time of one instruction on processor i, g is the BSP communication parameter and l the cost of synchronization [16,17] (review *proposition 1* of appendix A for the proof of this cost [10]).

Now we can easily create $Hist_i^{a_1}(R \bowtie S)$ by computing in parallel, on each processor i, the intersection of $Hist_i^{a_1}(R)$ and $Hist_i^{a_1}(S)$ in time of order:

$$Time_{phase2.b} = O\Big(\max_{i=1,\dots,p} \big(\gamma_i * \min(||Hist_i^{a_1}(R)||, ||Hist_i^{a_1}(S)||) \big) \Big).$$

While creating $Hist_i^{a_1}(R \bowtie S)$, we also store for each value $v \in Hist_i^{a_1}(R \bowtie S)$ an extra information $index(v) \in \{1, 2\}$ such that:

$$\begin{cases} index(v) = 1 & \text{if } Hist^{a_1}(R)(v) \geq f_o \text{ or } Hist^{a_1}(S)(v) \geq f_o \\ & \text{(i.e. values having high frequencies)} \\ index(v) = 2 & \text{elsewhere (i.e. values associated to low frequencies).} \end{cases}$$

The used threshold frequency is $f_o = p * log(p)$.

This information will be useful in the phase of the creation of communication templates. The total cost of this phase is the sum of $Time_{phase2.a}$ and $Time_{phase2.b}$. We recall that the size of a histogram is, in general, very small compared to the size of base relations.

Phase 3. Creating the communication template:
In homogeneous systems workload imbalance may be due to uneven distribution of data to be joined among the processors, while in heterogeneous systems it may be the result of allocating to processors an amount of tuples that is not proportional to actual capabilities of used machines [9].

So in order to achieve an accepted performance, the actual capacity of each machine must be taken into account while assigning the data or tasks to each processor. Another difficulty in such systems lies in the fact that available capacities of machines in multi-user systems may rapidly change after load assignment: the state of an overloaded processor may fastly become underloaded while computing the join operation and vice-versa. Thus to benefit from the processing power of such systems, we must not have idle processors while others are overloaded throughout all the join computation phase.

To this end, we use, as in *DFA-join* algorithm, a two-step (static then dynamic) load assignment approach which allows us to reduce the join processing time.

3.a. Static load assignment step:

In this step, we compute, in parallel on each processor i, the size of the join which may result from joining all tuples related to values $v \in Hist_i^{a1}(R \bowtie S)$. This is simply the sum of the frequencies $Hist_i^{a1}(R \bowtie S)(v)$ for all values v of the join attribute in $Hist_i^{a1}(R \bowtie S)$. This value is computed by a linear traversal of $Hist_i^{a1}(R \bowtie S)$ in time:

$$O\left(\max_{i=1,\ldots,p} \gamma_i * \|Hist_i^{a1}(R \bowtie S)\| \right).$$

After that all processors send the computed value to a designated head node which in its turn calculates the total number of tuples in $R \bowtie S$ (i.e. $\|R \bowtie S\|$) by computing the sum of all received values in time of order: $O(p * g + l)$.

Now the head node uses the value of $\|R \bowtie S\|$ and the information received earlier to assign to each processor i a join volume ($vol_i * \|R \bowtie S\|$) proportional to its resources where the value of vol_i is determined by the head node depending on the actual capacity of each processor i such that $\sum_{i=1}^{p} vol_i = 1$.

Finally, the head node sends to each processor i the value of $vol_i * \|R \bowtie S\|$ in time of order: $O(g * p + l)$.

The cost of this step is:

$$Time_{phase3.a} = O\left(\max_{i=1,\ldots,p} \gamma_i * \|Hist_i^{a1}(R \bowtie S)\| + p * g + l \right).$$

3.b. Communication templates creation step:

Communication templates are list of messages that constitute the relations redistribution. Owing to fact that values which could lead to AVS (those having high frequencies) are also those which may cause join product skew in "standard" hash join algorithms, we will create communication templates for only values v having high frequencies (i.e. $index(v) = 1$). Tuples associated to low frequencies (i.e. $index(v) = 2$) don't have effect neither on AVS nor on JPS. So these tuples will be simply hashed into buckets in their source processors using a hash function and their treatment will be postponed to the dynamic phase.

So first of all, $Hist_i^{a1}(R \bowtie S)$ is partitioned on each processor i into two sub-histograms: $Hist_i^{(1)}(R \bowtie S)$ and $Hist_i^{(2)}(R \bowtie S)$ such that: $v \in Hist_i^{(1)}(R \bowtie S)$ if $index(v) = 1$ and $v \in Hist_i^{(2)}(R \bowtie S)$ if $index(v) = 2$. This partitioning step is performed while computing $\sum_v Hist_i^{a1}(R \bowtie S)(v)$ in step 3.a in order to avoid reading the histogram two times.

We can start now by creating the communication templates which are computed, in a fist time, on each processor i for only values v in $Hist_i^{a1}(R \bowtie S)$ such that the total join size related to these values is inferior or equal to $vol_i * ||R \bowtie S||$ starting from the value that generates the highest join result and so on.

It is important to mention here that only tuples that effectively participate in the join result will be redistributed. These are the tuples of $\overline{R_i} = R_i \ltimes S$ and $\overline{S_i} = S_i \ltimes R$. These semi-joins are implicitly evaluated due to the fact that the communication templates are only created for values v that appear in join result (i.e. $v \in Hist^{a_1}(R \bowtie S)$).

In addition, the number of received tuples of R (resp. S) must not exceed $vol_i * ||\overline{R}||$ (resp. $vol_i * ||\overline{S}||$). For each value v in $Hist_i^{a_1}(R \bowtie S)$, processor i creates communicating messages $order\_to\_send(j, i, v)$ asking each processor j holding tuples of R or S with values v for the join attribute to send them to it.

If the processing capacity of a processor i doesn't allow it to compute the join result associated to all values v of the join attribute in $Hist_i^{a_1}(R \bowtie S)$[4], then it will not ask the source processors j holding the remaining values v to redistribute their associated tuples but to partition them into buckets using a hash function and save them locally for further join processing step in the dynamic phase. Hence it sends an $order\_to\_save(j, v)$ message for each processor j holding tuples having values v of the join attribute.

The maximal complexity of creating the communication templates is: $O\Big(\max_i \big(\omega_i * p * \gamma_i * ||Hist^{(1)}(R \bowtie S)|| \big) \Big)$, because each processor i is responsible of creating the communication template for approximately $\omega_i * ||Hist^{(1)}(R \bowtie S)||$ values of the join attribute and for a given value v at most $(p - 1)$ processors can send data.

After creating the communication templates, on each processor i, $order\_to\_send(j, i, .)$ and $order\_to\_save(j, .)$ messages are sent to their destination processors j when $j \neq i$ in time: $O\Big(\max_i \big(g * \omega_i * p * |Hist^{(1)}(R \bowtie S)| \big) + l \Big)$.

The total cost of this step is:

$$Time_{phase3.b} = O\Big(\max_i \big(\omega_i * p * (g * |Hist^{(1)}(R \bowtie S)| + \gamma_i * ||Hist^{(1)}(R \bowtie S)||) \big) + l \Big).$$

3.c. Task generation step:

After the communication templates creation, each processor i obeys the $order\_to\_send(., ., .)$ messages that it has just received to generate tasks to be executed on each processor. So it partitions the tuples that must be sent to each processor into multiple number of buckets greater than p using a hash function. This partition facilitates task reallocation in the join phase (phase 5) from overloaded to idle processors if a processor could not finish its assigned load before the others. After this partition step, each bucket is sent to its destination processor. The cost of this step is: $O\big(\max_{i=1,...,p} \gamma_i * (||R_i|| + ||S_i||) \big)$.

In addition, each processor i will partition tuples whose join attribute value is indicated in $order\_to\_save()$ messages into buckets using the same hash function on all the processors. However, these buckets will be kept for the moment in their source processors where their redistribution and join processing operations will be postponed till the dynamic phase.

During this step, local histogram of the join result, $(Hist^{b_1}(R \bowtie S))$, on attribute b_1 is created directly from $Hist^{a_1}(R_i)$ and $Hist^{a_1}(S_i)$ using Algorithm 2.

---

[4] This is the case if $vol_i * ||R \bowtie S|| < \sum_v Hist_i^{a_1}(R \bowtie S)(v)$ on processor i.

Algorithm 2. Join result histogram's creation algorithm on attribute b_1

Par   (on each node) $i \in [1, p]$ **do**
  ▷ $Hist^{b_1}(R_i \bowtie S_i)$=NULL (Create an empty B-tree to store histogram's entries)
  For each tuple t of each bucket of relation R_i **do**
    ▷ $freq1 = Hist^{a_1}(S)(t.a_1)$ (i.e. the frequency of the value of the attribute a_1 of t)
    If $(freq1 > 0)$ (i.e. tuple t will be present in $R \bowtie S$) **Then**
      ▷ $freq2 = Hist^{b_1}(R_i \bowtie S_i)(t.b_1)$
      If $(freq2 > 0)$ (i.e value $t.b_1$ is present in $Hist^{b_1}(R_i \bowtie S_i)$) **Then**
        ▷ Update $Hist^{b_1}(R_i \bowtie S_i)(t.b_1) = freq1 + freq2$
      Else
        ▷ Insert a new couple $(t.b_1, freq1)$ into the histogram $Hist^{b_1}(R_i \bowtie S_i)$
      End If
    End If
  End For
End Par

Owing to the fact that the access to the histogram (equivalent to a search in a B-tree) is performed in a constant time, the cost of the creation of the histogram of join result is : $O(max_{i=1,\ldots,p} \gamma_i * ||R_i||)$.
The global cost for this step:

$$Time_{phase3.c} = \left(max_{i=1,\ldots,p} \gamma_i * (||R_i|| + ||S_i||)\right).$$

Phase 4. Data redistribution:

According to communication templates, buckets are sent to their destination processors. It is important to mention here that only tuples of R and S that effectively participate in the join result will be redistributed. So each processor i receives a partition of R (resp. S) whose maximal size is $vol_i * ||\overline{R}||$ (resp. $vol_i * ||\overline{R}||$). Therefore, in this algorithm, communication cost is reduced to a minimum and the global cost of this phase is:

$$Time_{phase4} = O\left(g * max_i \left(vol_i * (|\overline{R}| + |\overline{S}|) + l\right)\right).$$

Phase 5. Join computation:

Buckets received by each processor are arranged in a queue. Each processor executes successively the join operation of its waiting buckets. The cost of this step of local join computation is:

$$Time_{local\_join} = O\left(\max_i \left(c^i_{r/w} * vol_i * (|\overline{R}| + |\overline{S}| + |\overline{R} \bowtie \overline{S}|)\right)\right).$$

If a processor finishes computing the join related to its local data and the overall join operation is not finished, it will send to the head node a message asking for more work. Hence, the head node will assign to this idle processor some of the buckets related to join attribute values that were not redistributed earlier in the static phase. However, if all these buckets are already treated, the head node checks the number of non treated buckets in the queue of the other processors and asks the processor that has the maximal number of non treated buckets to forward a part of them to the idle one. The number of sent buckets must respect the capacity of the idle processor.

And thus, the global cost of join computation of two relations R and S using *PDFA-Join* algorithm is:

$$
\begin{aligned}
Time_{PDFA-join} = O\Big(&\max_{i=1,\ldots,p} c^i_{r/w} * (|R_i| + |S_i|) + \max_{i=1,\ldots,p} \gamma_i * (||R_i|| + ||S_i||) + l \\
&+ \min\Big(\max_{i=1,\ldots,p} \omega_i * p * (g * |Hist^{a_1}(R)| + \gamma_i * ||Hist^{a_1}(R)||), \max_{i=1,\ldots,p} \omega_i * (g * |R| \\
&+ \gamma_i * ||R||)\Big) + \min\Big(\max_{i=1,\ldots,p} \omega_i * p * (g * |Hist^{a_1}(S)| + \gamma_i * ||Hist^{a_1}(S)||), \\
&\max_{i=1,\ldots,p} \omega_i * (g * |S| + \gamma_i * ||S||)\Big) + g * \max_{i=1,\ldots,p} (vol_i * (|\overline{R}| + |\overline{S}|) \\
&+ \max_{i=1,\ldots,p} (\omega_i * p * (g * |Hist^{(1)}(R \bowtie S)| + \gamma_i * ||Hist^{(1)}(R \bowtie S)||)) \\
&+ \max_{i=1,\ldots,p} (c^i_{r/w} * vol_i * (|\overline{R}| + |\overline{S}| + |\overline{R} \bowtie \overline{S}|)) \Big).
\end{aligned}
$$

Remark:
Sequential evaluation of the join of two relations R and S on processor i requires at least the following lower bound:

$$
bound_{inf_1} = \Omega\big(c^i_{r/w} * (|R| + |S| + |R \bowtie S|) + \gamma_i * (||R|| + ||S|| + ||R \bowtie S||)\big).
$$

Therefore, parallel join processing on p heterogeneous processors requires:

$$
bound_{inf_p} = \Omega\Big(\max_i \big(c^i_{r/w} * \omega_i * (|R| + |S| + |R \bowtie S|) + \gamma_i * \omega_i * (||R|| + ||S|| + ||R \bowtie S||)\big)\Big).
$$

PDFA-Join algorithm has optimal asymptotic complexity when:

$$
\max_{i=1,\ldots,p} (\omega_i * p * |Hist^{(1)}(R \bowtie S)|) \leq \max_i (c^i_{r/w} * \omega_i * \max(|R|, |S|, |R \bowtie S|)), \quad (1)
$$

this is due to the fact that other terms in $Time_{PDFA-join}$ are bounded by those of $bound_{inf_p}$.

Inequality (1) holds if the chosen threshold frequency f_o is greater than p (which is the case for our threshold frequency $f_o = p * log(p)$).

3.2 Discussion

To understand the whole mechanism of *PDFA-Join* algorithm, we compare existing approaches (based on hashing) to our pipelined join algorithm using different execution strategies to evaluate the multi-join query $Q = (R \bowtie_{a_1} S) \bowtie_{b_1} (U \bowtie_{a_2} V)$.

A *Full Parallel* execution of DFA-Join algorithm (i.e. a basic use of DFA-Join where we do not use pipelined parallelism) requires the evaluation of $Q_1 = (R \bowtie_{a_1} S)$ and $Q_2 = (U \bowtie_{a_2} V)$ on two disjoint set of processors, the join results of Q_1 and Q_2

are then stored on the disk. The join result of query Q_1 and Q_2 are read from disk to evaluate the final join query Q.

Existing approaches allowing pipelining first start by the evaluation of the join queries Q_1 and Q_2, and then each generated tuple in query Q_1 is immediately used to build the hash table. However the join result of query Q_2 is stored on the disk.

At the end of the execution of Q_1, the join result of query Q_2 is used to probe the hash table. This induces unnecessary disk input/output. Existing approaches require data redistribution of all intermediate join result (not only relevant tuples) this may induce high communication cost. Moreover data redistribution in these algorithms is based on hashing which make them very sensitive to data skew.

In PDFA-Join algorithm, we first compute in parallel the histograms of R and S on attribute a_1, and at the same time we compute the histograms of U and V on attribute a_2. As soon as these histograms are available, we generate the communication templates for Q_1 and Q_2 and by the way the histograms of the join results of Q_1 and Q_2 on attribute b_2 are also computed. Join histograms on attribute b_2 are used to create the communication templates for Q which makes it possible to immediately use the tuples generated by Q_1 and Q_2 to evaluate the final join query Q.

PDFA-Join algorithm achieves several enhancements compared to pipelined join algorithm presented in the literature: During the creation of communication templates, we create on the fly the histograms for the next join, limiting by the way the number of accesses to data (and to the disks). Moreover data redistribution is limited to only tuples participating effectively to join result, this reduces communication costs to a minimum. Dynamic data redistribution in *PDFA-Join* makes it insensitive to data skew while guaranteeing perfect load balance during all the stages of join computation.

PDFA-Join can be used in various parallel strategies, however in the parallel construction of the histograms for source relations, we can notice that the degree of parallelism might be limited by two factors : the total number of processors available, and the original distribution of data. A simultaneous construction of two histograms on the same processor (which occurs when two relations are distributed, at least partially, over the same processors) would not be really interesting compared to a sequential construction. This intra-processor parallelism does not bring acceleration, but should not induce noticeable slowdown : histograms are generally small, and having several histograms in memory would not necessitate swapping. On the other hand, as relations use to be much bigger than the available memory, we have to access them by blocks. As a consequence, accessing one or several relations does not really matter. Our pipeline strategy will really be efficient if different join operators are executed on disjoint (or at least partially disjoint) sets of processors. This brings us to limit the number of simultaneous builds. As a consequence, we have to segment our query trees, similarly to segmented right-deep trees, each segment (i.e. a set of successive joins) being started when the former is over. Once the histograms are produced for both tables, we can compute the communication template, then distribute data, and finally compute the join. Unfortunately, the computation of the communication template is the implicit barrier within the execution flow, that prohibits the use of long pipeline chains.

4   Conclusions

In this paper, we presented *PDFA-Join* a pipelined parallel join algorithm based on a dynamic data redistribution. We showed that it can be applied efficiently in various parallel execution strategies offering flexible resource allocation and reducing disks input/output of intermediate join result in the evaluation of multi-join queries. This algorithm achieves several enhancements compared to solutions suggested in the literature by reducing communication costs to only relevant tuples while guaranteeing perfect balancing properties on heterogeneous multi-processors shared nothing architectures even for highly skewed data.

The BSP cost analysis showed that the overhead related to histogram management remains very small compared to the gain it provides to avoid the effect of load imbalance due to data skew, and to reduce the communication costs due to the redistribution of the intermediate results which can lead to a significant degradation of the performance.

Our experience with the BSP cost model and the tests presented in our previous papers [4,3,10] prove the effectiveness of our approach compared to standard hash-join pipelined algorithms.

References

1. Bamha, M.: An optimal and skew-insensitive join and multi-join algorithm for ditributed architectures. In: Andersen, K.V., Debenham, J., Wagner, R. (eds.) DEXA 2005. LNCS, vol. 3588, pp. 616–625. Springer, Heidelberg (2005)
2. Bamha, M., Exbrayat, M.: Pipelining a skew-insensitive parallel join algorithm. Parallel Processing Letters 13(3), 317–328 (2003)
3. Bamha, M., Hains, G.: A skew insensitive algorithm for join and multi-join operation on Shared Nothing machines. In: Ibrahim, M., Küng, J., Revell, N. (eds.) DEXA 2000. LNCS, vol. 1873, pp. 644–653. Springer, Heidelberg (2000)
4. Bamha, M., Hains, G.: A frequency adaptive join algorithm for Shared Nothing machines. Journal of Parallel and Distributed Computing Practices (PDCP) 3(3), 333–345 (1999); Appears also in: Columbus, F. Progress in Computer Research, vol. II. Nova Science Publishers (2001)
5. Chen, M.-S., Lo, M.L., Yu, P.S., Young, H.C.: Using segmented right-deep trees for the execution of pipelined hash joins. In: Yuan, L.-Y. (ed.) Very Large Data Bases: VLDB 1992, Proceedings of the 18th International Conference on Very Large Data Bases, Vancouver, Canada, August 23–27, pp. 15–26. Morgan Kaufmann Publishers, Los Altos (1992)
6. Chen, M.-S., Yu, P.S., Wu, K.-L.: Scheduling and processor allocation for the execution of multi-join queries. In: International Conference on Data Engineering, pp. 58–67. IEEE Computer Society Press, Los Alamos (1992)
7. Datta, A., Moon, B., Thomas, H.: A case for parallelism in datawarehousing and OLAP. In: Ninth International Workshop on Database and Expert Systems Applications, DEXA 1998, pp. 226–231. IEEE Computer Society, Vienna (1998)
8. DeWitt, D.J., Naughton, J.F., Schneider, D.A., Seshadri, S.: Practical Skew Handling in Parallel Joins. In: Proceedings of the 18th VLDB Conference, Vancouver, British Columbia, Canada, pp. 27–40 (1992)
9. Anastasios Gounaris: Resource aware query processing on the grid. Thesis report, University of Manchester, Faculty of Engineering and Physical Sciences (2005)

10. Hassan, M.A.H., Bamha, M.: Dynamic data redistribution for join queries on heterogeneous shared nothing architecture. Technical Report 2, LIFO, Université d'Orléans, France (March 2008)
11. Hua, K.A., Lee, C.: Handling data skew in multiprocessor database computers using partition tuning. In: Lohman, G.M., Sernadas, A., Camps, R. (eds.) Proc. of the 17th International Conference on Very Large Data Bases, Barcelona, Catalonia, Spain, pp. 525–535. Morgan Kaufmann, San Francisco (1991)
12. Liu, B., Rundensteiner, E.A.: Revisiting pipelined parallelism in multi-join query processing. In: VLDB 2005: Proceedings of the 31st international conference on Very large data bases, pp. 829–840. VLDB Endowment (2005)
13. Lu, H., Ooi, B.-C., Tan, K.-L.: Query Processing in Parallel Relational Database Systems. IEEE Computer Society Press, Los Alamos (1994)
14. Mourad, A.N., Morris, R.J.T., Swami, A., Young, H.C.: Limits of parallelism in hash join algorithms. Performance evaluation 20(1/3), 301–316 (1994)
15. Rahm, E.: Dynamic load balancing in parallel database systems. In: Fraigniaud, P., et al. (eds.) Euro-Par 1996. LNCS, vol. 1123. Springer, Heidelberg (1996)
16. Skillicorn, D.B., Hill, J.M.D., McColl, W.F.: Questions and Answers about BSP. Scientific Programming 6(3), 249–274 (1997)
17. Valiant, L.G.: A bridging model for parallel computation. Communications of the ACM 33(8), 103–111 (1990)
18. Wilschut, A.N., Apers, P.M.G.: Dataflow query execution in a parallel main-memory environment. In: Parallel and Distributed Information Systems (PDIS 1991), pp. 68–77. IEEE Computer Society Press, Los Alamits (1991)
19. Wilschut, A.N., Flokstra, J., Apers, P.M.G.: Parallel evaluation of multi-join queries. Proceedings of the ACM-SIGMOD 24(2), 115–126 (1995)

Declarative Business Process Modelling
and the Generation of ERP Systems

Nicholas Poul Schultz-Møller, Christian Hølmer, and Michael R. Hansen

Department of Informatics and Mathematical Modelling
Technical University of Denmark, 2800 Kgs., Lyngby, Denmark
nicholassm@gmail.com, hoelmer@upsido.com, mrh@imm.dtu.dk

Abstract. We present an approach to the construction of *Enterprise Resource Planning (ERP)* Systems, which is based on the *Resources, Events and Agents (REA)* ontology. This framework deals with processes involving exchange and flow of resources in a declarative, graphically-based manner describing what the major entities are rather than how they engage in computations. We show how to develop a domain-specific language on the basis of REA, and a tool which automatically can generate running web-applications. A main contribution is a proof-of-concept showing that business-domain experts can generate their own applications without worrying about implementation details.

In order to have a well-defined domain-specific language, a formal model of REA has been developed using the specification language Object-Z and this led to clarifications as well as the introduction of new concepts. The compiler for our language is written in Objective CAML and as implementation platform we used Ruby on Rails. Our aim here is to give an overview of whole construction of a running application from a REA specification and to illustrate the adequacy of the development process.

Keywords: Enterprise Resource Planning systems, REA, Declarative domain specific language, Web-applications.

1 Introduction

In this paper we present an approach to the construction of *Enterprise Resource Planning (ERP)* Systems, which is based on the *Resources, Events and Agents (REA)* ontology [1,2,3]. REA is a domain-specific conceptual framework which has its use, for example, in the design of accounting and enterprise information systems. Though this framework deals with processes involving exchange and flow of resources, the conceptual models have high-level graphical representations describing what the major entities are rather than how they engage in computations.

We show how to develop a declarative, domain-specific language on the basis of the REA ontology and a tool which can generate running web-applications. A main contribution is a proof-of-concept showing that business-domain experts can, in principle, using a REA-based domain-specific language, generate their own applications without worrying about implementation details. This has a clear advantage compared to current domain-specific languages, which have a much more imperative flavor (e.g. Microsoft Dynamics' C/AL language [4]).

J. Cordeiro et al. (Eds.): ICSOFT 2008, CCIS 47, pp. 134–146, 2009.

The REA ontology is not formalized and the informal explanations of REA are in many ways underspecified. To get a well-defined domain-specific language and an associated tool we develop a formal model of REA using the specification language Object-Z [5,6]. This object-oriented, formal specification language proved well-suited for our purpose as the underlying computational model should describe the changing state of resources. The formalization process led to several clarifications as well as the introduction of new concepts.

The formal model exhibits in a succinct way fundamental properties of resources, events and agents, which any implementation should satisfy. In order to achieve this for our implementation we used Objective CAML [7], which is a functional programming language with object-oriented features. The Object-Z model could be represented in Objective CAML in a direct manner, and the functional part of the programming language proved very useful for handling the formal properties of the model. This close correspondence made the development time shorter and enhanced the confidence in the correctness of the implementation. As implementation platform for the web application we used Ruby on Rails [8].

The aim of this paper is to give an overview of the whole construction of a running application on the basis of a REA specification, and to justify the thesis that a running application can be automatically generated from a declarative, REA-based specification. We will sketch the flavor of the Object-Z model and how it relates to the Objective CAML implementation, to emphasize the adequacy of this approach. Furthermore, we will hint at the use of the implementation platform Ruby on Rails.

Current ERP systems try to satisfy the majority of their customers' requirements and customize the last 15-20% by using domain specific languages such as C/AL. These languages are usually imperative and are used for purposes such as implementation of windows, dialogs, algorithms, etc. As a result many resources are spent on customization and the extensions are unfortunately tightly coupled to the basic ERP system. This prohibits the existing systems from benefitting from new advantages in language technologies and architectural designs.

We suggest another approach for creating customized ERP systems, where the fundamental idea is to decouple the description of the business processes from design and implementation issues [9]. By declaring a customer's business processes and then generate the ERP system, it will be possible to consistently extend the functionality as business processes develop. This enables transparent commissioning of new technologies.

Another declarative approach to the modelling and analysis of ERP systems is described in [10,11], which investigates two ways of dealing with value-added tax. One is based on OWL with related tools, and the other was based on the Configit Product Modelling Language (CPML) [12], which consists of variables with finite domains, relations and functions, and a corresponding tool which is BDD-based. This work, which aimed at analysis rather than application generation, showed that OWL had severe limitations due to lack of support of integer arithmetic, and CPML did not provide the right level of abstraction.

This paper, which is based on [13], is organized as follows: In the next section we give an introduction to REA on the basis of a simple example. Thereafter, in Section 3,

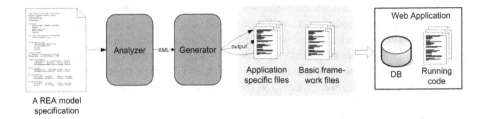

Fig. 1. From REA model specification to web application

we introduce a formal Object-Z model for REA. In Section 4 and 5 we describe the domain specific language, how it is analyzed and how web-applications are generated automatically. The overall idea is illustrated on Fig. 1. The last section contains a discussion of our work. In the paper we will sketch the main ideas only, for further details we refer to [13].

2   An Introduction to REA

The ontology *Resources, Events and Agents* (*REA*) [1,2,3] can be used for describing business processes, such as the *exchange* of resources between economic agents, and the *conversion* (or *transformations*) of resources. REA is an ontological framework where the models typically are presented graphically like, for example, in Fig. 2, which describes an exchange process for a fictive company "Paradice", that produces and sells ice cream.

To produce ice cream it is necessary to purchase cream from a retailer and the figure describes that the resource Cream is exchanged in a *duality* for Capital in the events Cream Delivery and Cash Disbursement, respectively. The relationships show that the

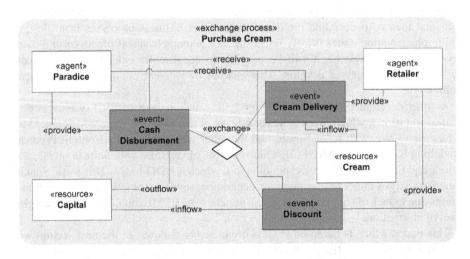

Fig. 2. Modelling the purchase of cream in REA

agent Paradice *provides* the capital and *receives* cream. The *inflow* and *outflow* relationships between an event and a resource describe whether or not *value* flows into (an *increment event*) or out of the enterprise (a *decrement event*). Here the enterprise's capital resources decrease their value, while its cream resources increase their value – simply because the amount of each resource changes. Thus, Cash Disbursement is a decrement event and Cream Delivery is an increment event. The figure also contains an *optional* Discount event modelling that a retailer might give a discount on a purchase, but it is not possible to infer from the diagram that this event is optional.

Using REA, one can model a business process by defining distinct events involving agents and resources. REA has several rules that must apply to a model for it to be valid, for example: (1) Any resource must be part of at least one increment and one decrement event, and (2) Any increment event must be in a duality relationship with at least one decrement event and vice versa. Rule (1) ensures that no resource "piles up" or is sold but never purchased, and rule (2) ensures that it is always possible to track what other resources a specific resource was exchanged for. The REA ontology does, however, not have a precise definition in a formal sense. This is a problem when a system is implemented on the basis of an ontology, as the developer may take wrong decisions which can lead to an inconsistent design.

Example problems in REA are: (1) It is unclear how to distinguish two resources of the same type from each other as distinguishable resources have no notion of identity and (2) it is not possible to distinguish events that must occur in a business process and events which are optional (as the Discount event). These and other problems have been considered and solved in [13]. We will elaborate on that in the next section.

3   An Object-Z Model

We will now sketch a formal model for REA. For a complete account we refer to [13]. The object-oriented specification language Object-Z [5,6] was chosen for the formalization as business processes to a large extend concern the changing states and exchanges of resources. The model is divided in two parts: A *meta-model* defining the concepts of the REA ontology and their properties, and a *runtime model* defining the dynamic behavior of applications. We will not give a dedicated introduction to Object-Z, but just give brief informal explanations to the shown specifications.

3.1   The REA Meta-model

The REA meta-model contains an Object-Z class for each entity in the REA ontology and a formalization of the properties all instances of these classes must satisfy. We first define *free types* in the meta-model for names and field types:

$$NAME ::= nil \mid charseq \langle\!\langle \text{seq}_1 \, CHAR \rangle\!\rangle$$
$$FTYPE ::= sType \mid iType \mid fType \mid bType$$

where *NAME* is either undefined or a sequence of characters, and *FTYPE*, a field type, is one of the following **s**tring, **i**nteger, **f**loat and **B**oolean. Fields are used for augmenting the entities – e.g. an address field on a retailer agent entity. Furthermore, fields are modelled as partial functions from names to field types:

$$FIELDS == NAME \nrightarrow FTYPE$$

The entities in the meta-model share many properties and these are specified in the *BasicEntity* class:

┌─ *BasicEntity* ───
│ ┌──
│ │ $name : NAME, fields : FIELDS$
│ │
│ │ ┌─ *INIT* ──
│ │ │ $name = nil, fields = \varnothing$
│ │
│ │ ┌─ *InitData* ──────────────────────────────────────
│ │ │ $\Delta(name, fields)$
│ │ │ $name? : NAME, fields? : FIELDS$
│ │ │
│ │ │ $name = nil \wedge name? \neq nil$
│ │ │ $name' = name? \wedge fields' = fields?$
│ └──

The *precondition* of *InitData* ensures that an entity can only set data once (and in that respect works as a constructor) and that each entity is named.

A class for resources is defined by extending *BasicEntity*:

┌─ *Resource* ──
│ ⌈ $(name, fields, atomic, InitData)$
│ *BasicEntity*
│ ┌──
│ │ $atomic : \mathbb{B}, key : NAME$
│ │
│ │ $atomic \Leftrightarrow key \in (dom\, fields)$
│ │ ┌─ *INIT* ──
│ │ │ $atomic = false, key = nil$
│ │
│ │ ...
│ └──

where the variable *atomic* specifies whether a real world resource object is distinguishable from another of the same type. If so, there must be a way to identify a resource uniquely in the Enterprise Information System. This solution to the problem of distinguishing two resources from each other was chosen to adapt the key concept from databases.

A class for agents is defined as follows:

Agent
$\lceil$ (name, fields, InitData)
BasicEntity

For the definition of the *Event* class we introduce two additional free types *STOCK-FLOW* (the relationship between an event and a resource) and *MULTIPLICITY* (the number of occurrences of an event in a process):

$$STOCKFLOW ::= none \mid flow \mid produce \mid use \mid consume$$

$$MULTIPLICITY ::= \langle\!\langle \mathbb{N}_0 \rangle\!\rangle \mid infty$$

Observe that inflow and outflow have been generalized to *flow*, as the direction can be inferred from the provide/receive relationship. E.g., an agent providing in an event is an outflow of a resource under the possession of that agent.

A class for events is defined in Fig. 3, where we, for example, observe that an exchange event can have flow only as stock flow type. Several *Events* constitute a process as sketched in Fig. 4, with the most interesting invariant being [a], which states that one of the agents, called the *enterprise-agent*, must be receiver or provider in all events. The enterprise-agent is a new central concept that represents the enterprise described in a model. As will be apparent below, our system can automatically infer[1] which agent in a consistent model is the enterprise-agent.

All instances in a model are collected in the *MetaModel* class shown in Fig. 5. The formulas [b] and [c] state that all agents and resources must be part of at least one event. Formula [d] expresses that the enterprise-agent must be either provider or receiver in any exchange events. Hence, it is possible to infer the enterprise-agent. Formula [e] is central to the model [13]. In the meta-model a flow of resources is interpreted as resources that either leave or enter the enterprise – not as a flow of value. Thus, any resource must both leave the enterprise and enter the system in a decrement and an increment event, respectively.

3.2 The Runtime Model

The runtime model addresses the dynamic aspects of a REA-model specification. We will just mention two aspects, referring to [13] for further details:

– An *execution* of a REA-specification is a (possibly infinite) sequence of *states*, where a state (among other things) records the *available* resources in the system. The runtime model defines all legal executions of a system, where, for example, a decrement event can happen in the current state only if there are sufficient available resources for that event.

---

[1] Very unrealistic business models can be constructed which prohibit automatic inference. In these cases the user is prompted for selection of one agent.

$\ulcorner$ *Event* _____
$\ulcorner$ (*name, fields, provider, receiver, stockFlowType,*
       resource, minOccurs, maxOccurs, InitData,
       IsExchange, IsConversion)

BasicEntity
_____

   provider, receiver : *Entity* $\cup$ *Agent*
   stockFlowType : *STOCKFLOW*
   resource : *Entity* $\cup$ *Resource*
   minOccurs, maxOccurs : *MULTIPLICITY*
   _____
   $\neg(maxOccurs = 0) \wedge minOccurs \neq infty$

   $maxOccurs \in \mathbb{N} \Rightarrow minOccurs \leq maxOccurs$

$\ulcorner$ *INIT* _____
   provider, receiver = *Entity.INIT*
   stockFlowType = *none*
   resource = *Entity.INIT*
   minOccurs, maxOccurs = 0

...

$\ulcorner$ *IsExchange* _____
   stockFlowType = *flow*

$\ulcorner$ *IsConversion* _____
   stockFlowType $\in$ {*use, consume, produce*}

Fig. 3. The Object-Z class *Event*

 – An agent is a person or organization, but e.g. a person may *act* as different agents in
   different events, e.g. as an employee in one event and as a customer in another. To
   cope with this, a new concept *legal entity* is introduced, where a legal entity (person
   or organization) is associated with a set of agents that the legal entity can act as in
   a given event.

4   A Domain Specific Language

A domain specific language for specifying REA models has been designed so that business domain experts can define business processes. The concepts of the Meta-Model
are directly reflected in the language as can be seen from the abstract syntax presented
in Fig. 6.

   As an example of the language and its semantics, consider Fig. 7, which gives the
specification corresponding to Fig. 2.

$\boxed{\begin{array}{l} \rule{0pt}{0pt}\text{\emph{Process}} \\[2pt] \restriction (name, dualities, INIT, InitData) \\[6pt] \hline name : NAME, dualities : \mathbb{F}\ Event \\[6pt] \exists\, e_1 : dualities \bullet e_1.IsExchange \Rightarrow \\[6pt] \hline \text{\emph{INIT}} \\ name = nil, dualities = \varnothing \\[6pt] \hline \dots \end{array}}$

Fig. 4. The Object-Z class *Process*

$\boxed{\begin{array}{l} \text{\emph{MetaModel}} \\[2pt] \restriction (enterpriseAgent, resources, processes, \\ \qquad agents, events, InitData, \\ \qquad SpecifyEnterpriseAgent, IsValid) \\[6pt] \hline enterpriseAgent : Entity \cup Agent \\ resources : \mathbb{F}\ Resource,\ agents : \mathbb{F}\ Agent \\ events : \mathbb{F}\ Event,\ processes : \mathbb{F}\ Process \\ \dots \\[6pt] \hline \{p : Process \mid p \in processes \bullet p.dualities\} \\ \qquad \text{partitions } events \\[4pt] \forall\, a : agents;\ \exists\, e : events \bullet \\ \qquad e.receiver = a \vee e.provider = a \\[6pt] \hline \text{\emph{INIT}} \\ enterpriseAgent = Entity.INIT \\ resources = \varnothing,\ agents = \varnothing \\ events = \varnothing,\ processes = \varnothing \\ \dots \\[6pt] \hline \dots \end{array}}$

Fig. 5. The Object-Z class *MetaModel*

The types of fields are declared in the first line. Two agents, Paradice and Retailer, are declared in the second line and two resources, Capital and Cream with their fields, are declared in line 3. These resources are non-atomic as no field is marked as key with a '*' following the field name. Then the process, BuyCream, is declared. It consists of three events. The declaration of the first event, CashDisbursement, can be read as: "CashDisbursement must occur exactly once in a BuyCream process and is a flow of

specification ::= decl$^+$
decl   ::= **agent** (id [fields])$^+$ | **resource** (id [key_fields])$^+$
            | **resourcetype** id [key_fields] { (id [key_fields])$^+$ }
            | **process** id { event_decl$^+$ } | type id$^+$
field   ::= [id$^+$]
key_ field   ::= [(id* | id) id*]
type   ::= **string** | **int** | **float** | **boolean**
event_decl ::= id [fields] : [multiplicity] flow_type id (id $->$ id)
multiplicity ::= number | number **..** number | number **..** *
flow_type   ::= **flow** | **use** | **consume** | **produce**

Fig. 6. Abstract syntax for the domain specific language, where 'number' ranges over natural numbers and 'id' ranges non-empty sequences of alphanumerical characters starting with a letter

```
1 string currency, description, address, unit, expireDate
2 agent Paradise, Retailer [address]
3 resource Capital [currency],Cream[description,expireDate,unit]
4 process BuyCream
5    { CashDisbursement: 1 flow Capital(Paradice -> Retailer)
6      CreamDelivery:    1 flow Cream  (Retailer -> Paradice)
7      Discount:       0..1 flow Capital(Retailer -> Paradice) }
```

Fig. 7. Specification corresponding to Figure 2

Capital from Paradice to a Retailer." This example also shows an optional Discount event, where the multiplicity is 0..1. The upper bound of the multiplicity can be '*' which means an infinite number of the event can occur in the business process. The full specification has six other processes: OwnerShip, MachineOwnerShip, BuyElectricity, AquireLabour, MakeIcecream and SellIcecream. These are omitted here for brevity as they have a similar structure.

4.1 Language Implementation

The language is implemented using Objective Caml [7,14], which is a programming language supporting functional as well as object-oriented programming. This makes it particularly suitable for constructing a programming model on the basis of a REA specification – the classes in the program are related to the classes in the Object-Z meta-model in a very direct manner. Fig. 8 provides an example by showing the OCaml implementation of the initData method in the Object-Z class Process (Fig. 4). The implementation uses high-order functions (Event_set.for_all and Event_set. exists) to directly express the required properties of the Meta-Model.

If the specification is consistent, i.e. no exception is thrown, then an XML representation is generated. In the next section it is sketched how a web-application can be generated on the basis of that XML representation.

```
 1 method initData ( nameIN, dualitiesIN ) =
 2 (* Check of class invariant. *)
 3 if Event_set.exists (fun e1 -> e1#isExchange =>
 4    Event_set.for_all (fun e2 ->
 5       e1#provider=e2#provider || e1#provider=e2#receiver
 6    ) dualitiesIN
 7    ||
 8    Event_set.for_all (fun e2 ->
 9       e1#receiver=e2#provider || e1#receiver=e2#receiver
10    ) dualitiesIN
11 ) dualitiesIN then ()
12 else modelError(( (* Raises an exception *)
13    "The events in the process "^(toStr nameIN)^" do not all"^
14    "have the same agent as either receiver or provider."),b#pos);
15 ...
16 name        <- nameIN
17 dualities   <- dualitiesIN
```

Fig. 8. Objective Caml implementation of the class invariant as specified in the Object-Z class Process. The class has two fields name and dualities. The operator => is the boolean implies operator.

5 Automated Generation of Applications

Having a consistent REA model, given now as an XML document, we can generate an application automatically. We have chosen to implement the application in the web-framework, Ruby on Rails [15,8], which let us focus on developing the features of the application instead of Model-View-Control plumbing.

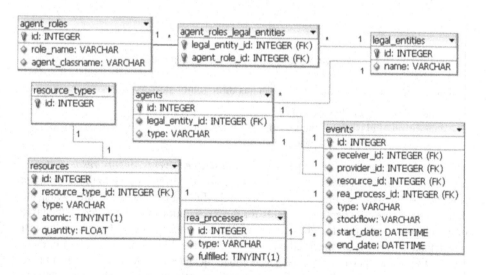

Fig. 9. Database overview

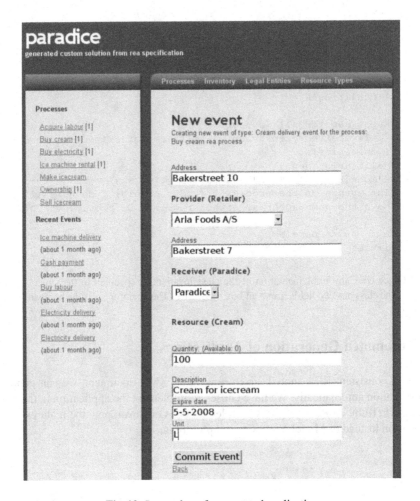

Fig. 10. Screenshot of a generated application

The generation of the implementation has three major steps:

- generation of a database schema,
- generation of classes for the specified entities, and
- generation of a runtime storage system.

The main idea is that each of the REA entities: *resources, events, agents, resource types, legal entities and processes* have a table with relations corresponding to those from the runtime model. We use the technique of Single Table Inheritance, so that when the database structure is generated all fields for the REA entities will be added to the table (e.g. field_description, field_serial_number). This gives us the flexibility of having just one database table for all specified types of the individual entities, but still letting them have individual properties. Especially when loading and saving it is useful to be able to treat them the same way. Note however that a different approach is needed for real-life ERP systems in order for the system to scale. Fig. 9 gives an overview of the database.

We also generate a Ruby class for each of the Single Table Inheritance types (e.g. `CreamDeliveryEvent` a subclass of `Event`), which will refer to the entries in a database table with the same entity name.

The generated entities constitute a running web-application and Fig. 10 shows the web-interface generated from the Paradice example of which a part of the specification is shown in Fig. 7. A user can register an event, e.g. a purchase of cream, by simply clicking on the relevant business process, click on the event and enter the details as shown on Fig. 10. If the resources are available in the current state of the database and the entered data are correct (e.g. type check of input fields) then the event is registered and the database will change state as described in the runtime model given in Object-Z.

The implementations of these events are derived from the specification of the runtime model; but we will not give further details here.

Extensions to the Automatic Approach

All enterprises are ever changing and the possibility of extending a running application would be beneficial. The problem with this is that all the recorded data must stay intact and the current business processes must therefore not be changed. The simplest solution was therefore only to allow addition of new business processes and prohibit changes in the current entities in the application. In this way enterprises can adapt to changes in their business by marking old entities as outdated and use newer ones instead.

6 Summary

We have justified the thesis of this paper, i.e. that business applications can be developed from a domain specific language based on the REA ontology. The justification is based on a prototype system which can generate web-based applications from REA specifications. So far the fundamental entities of the REA ontology, Resources, Events and Agents, have been considered and formalized, but we see no principal difficulties in extending the system to the full ontology. The prototype system is for proof-of-concept only. Currently it is not possible to express constraints on the temporal ordering of events. E.g. a business-domain expert may want to specify that a Cash Disbursement should only take place after a Cream Delivery. However, this can also be solved in a declarative way by defining a partial ordering on events. In the specification this could be expressed as `process BuyCream{...}[CashDisbursement<CreamDelivery]`, where $e_1 < e_2$ means that e_2 causally depends on e_1.

In order to get a fully functional ERP system many other aspects need to be addressed, for example workflow, access control, the possibility of creating custom reports etc. Integrating these aspects into the framework is future work.

Our method is the following: first a formal model of REA is expressed in Object-Z. This clarified many vague points about REA, and forms the basis for the definition of a domain specific language for declaring business processes and its implementation. The implementation consists of an analyzer and a compiler, both written in Objective Caml. Objective Caml is a functional programming language with object oriented features, and with this language we can express the Object-Z model for REA (classes, operations and rules) in a straightforward way. The analyzer performs the consistency checks as stated in the formal model, and the compiler generates a web-based application satisfying the rules of REA. As implementation framework Ruby on Rails was used.

Acknowledgements. We are grateful for comments and suggestions from Ken Friis Larsen. This work is partially funded by ArtistDesign (FP7 NoE no 214373), MoDES (Danish Research Council 2106-05-0022) and the Danish National Advanced Technology Foundation under project DaNES.

References

1. Geerts, G.L., McCarthy, W.E.: The ontological foundation of rea enterprise information systems (2000)
2. Hruby, P.: Model-Driven Design Using Business Patterns. Springer, Heidelberg (2006)
3. McCarthy, W.E.: The rea accounting model: A generalized framework for accounting systems in a shared data environment. The Accounting Review (3) (July 1982)
4. Studebaker, D.: Programming Microsoft Dynamics NAV. Packt Publishing (2007)
5. Smith, G.: The Object-Z Specification Language. Kluwer Academic Publishers, University of Queensland, Australia (2000)
6. Spivey, J.M.: The Z Notation, A Reference Manual, 2nd edn. Prentice Hall International (UK) Ltd., Englewood Cliffs (2001)
7. Smith, J.B.: Practical Ocaml. APress (2006)
8. Thomas, D., Hansson, D.H.: Agile Web Development with Rails, 2nd edn. Pragmatic Bookshelf (2006)
9. Hansen, M.R., Hansen, B.S., Lucas, P., van Emde Boas, P.: Integrating constraint languages and relational databases. Computer Languages 14(2), 63–82 (1989)
10. Nielsen, M.I., Simonsen, J.G., Larsen, K.F.: Requirements for logical models for value-added tax legislation. Presented at Logic Programming and Automated Reasoning 2008, Doha 2008. To appear in an EasyChair Collection volume (2008)
11. Nielsen, M.I.: Logical models for value-added tax legislation. Master's thesis, Department of Computer Science, University of Copenhagen, Njalsgade 128-132 DK-2300 Copenhagen S (2008)
12. ConfigIt: Product modeler 4.1 limited edition, http://www.configit.com
13. Schultz-Møller, N.P., Hølmer, C.: Bachelor Thesis: Tool Support for Business Processes - REAML, The REA Language and Metamodel (2007)
14. May, J.H.: Introduction to the Objective Caml Programming Language (2006)
15. Thomas, D., Fowler, C., Hunt, A.: Programming Ruby, 2nd edn. Pragmatic Bookshelf (2004)

Single Vector Large Data Cardinality Structure to Handle Compressed Database in a Distributed Environment

B.M. Monjurul Alom, Frans Henskens, and Michael Hannaford

School of Electrical Engineering and Computer Science
University of Newcastle, NSW 2308, Australia
Bm.Alom@studentmail.newcastle.edu.au
{Frans.Henskens,Michael.Hannaford}@newcastle.edu.au

Abstract. Loss-less data compression is attractive in database systems as it may facilitate query performance improvement and storage reduction. Although there are many compression techniques that handle the whole database in main memory, problems arise when the amount of data increases gradually over time, and also when the data has high cardinality. Management of a rapidly evolving large volume of data in a scalable way is very challenging. This paper describes a disk based single vector large data cardinality approach, incorporating data compression in a distributed environment. The approach provides substantial storage performance improvement compared to other high performance database systems. The presented compressed database structure provides direct addressability in a distributed environment, thereby reducing retrieval latency when handling large volumes of data.

Keywords: Fragment, Main Memory, Compression, Vector, Cardinality.

1 Introduction

In main memory database systems data resides permanently in main physical memory, whereas in a conventional database system it may be disk resident [1]. Conventional database systems cache data in main memory for access; in main memory database systems data may have a backup copy on disk. In both cases, therefore, a given object can have copies both in memory and on disk. The key difference is that in main memory databases the primary copy lives permanently in memory. Main memory is normally directly accessible and volatile while disks are not. The layout of data on a disk is much more critical than the layout of data in main memory, since sequential access to a disk is faster than random access [1]. The main pitfall of main memory databases is that they cannot handle very large amounts of data because they are fully dependent on main memory. This can be alleviated somewhat, for example the HIBASE compression technique [2] is main memory based and applicable to low cardinality of domain values. In this paper we present a single vector large data cardinality structure (SVLDCS) that is disk based and supports large databases as well as

J. Cordeiro et al. (Eds.): ICSOFT 2008, CCIS 47, pp. 147–160, 2009.

high cardinality of domain values with the facility to access compressed data in distributed environments. Portions of the compressed database structure are available in main memory on the basis of the query demand from different sites. The main copy of the domain dictionary is stored permanently on the disk and a back up copy is available in the main memory. This structure is used to handle large scale of tuples and attributes while providing a level of storage performance comparable to conventional database systems. This structure may be easily used in areas where database is often typically only queried, for example for analytical processing, data warehousing and data mining applications.

The remainder of this paper is organized as follows: Related work is described in section 2. The existing HIBASE method is presented in section 2.1 - our method is an extension of this architecture. The (SVLDCS) single vector large data cardinality structure is described in 3. Section 4 and 5 present the search technique and analysis of storage capacity respectively. The paper concludes with a discussion and final remarks in section 6.

2   Related Work

The HIBASE architecture [2] defines a way of representing a dictionary based compression technique for relational databases that are fully main memory based. This structure is designed for low cardinality of the domain dictionaries, with an architecture that replaces code rather than the original data values in tuples. The main pitfall of this method is that the structure cannot handle large databases because of full dependency on main memory in combination with the limitations of large memory spaces. Investigation of main memory database systems is well described in [1], with a major focus on fidelity of main memory content compared to conventional disk based database systems. Memory resident database systems (MMDB's) store their data in main physical memory, providing high data access speeds and direct accessibility. As semiconductor memory becomes less expensive, it is increasingly feasible to store databases in main memory and for MMDB's to become a reality.

The unique graph-based data structure called DBGraph may be used as a database representation tool that fully exploits the direct access capability of main memory. Additionally, the rapidly decreasing cost of RAM makes main memory database systems a cost effective solution to high performance data management [3]. This overcomes the problem that disk-based database systems have their performance limited by I/O.

A compressed 1-ary vertical representation is used to represent high dimensional sparsely populated data where database size grows linearly [4]. Queries can be processed on the compressed form without decompression; decompression is done only when the result is necessary. Different kinds of problem, such as access control and transaction management, may apply to distributed and replicated data in distributed database systems (DDBMS) [5]. Oracle, a leading commercial DBMS, defines a way to maintain consistent state for the database using a distributed two phase commit protocol. [5] addresses some issues such as advantage, disadvantage, and system failure in distributed database systems. Since organizations tend to be geographically dispersed, a DDMBS fits the organizational structure better than traditional centralized DBMS.

Advantages of DDBMS include that failure of a server at one site will not necessarily render the distributed database system inaccessible. A general architecture for archiving and retrieving real-time, scientific data is described in [6]. The basis of the architecture is a data warehouse that stores metadata on the raw data to allow for its efficient retrieval. A transparent data distribution system uses the data warehouse to dynamically distribute the data across multiple machines.

A single dictionary based compression technique to manage large scale databases is described by Oracle corporation [7]. The authors also address an innovative table compression technique that is very attractive for large relational data warehouses. This technique is used to compress and partition tables. The status of a table can be changed from compressed to non-compressed at any time by simply adding the keyword COMPRESS to the table's meta-data.

The LH*$_{RS}$ scheme defines a way of storing available distributed data [8]. This system includes distributed data structures [SDDS] that are intended for computers over fast networks, usually local networks. The architecture is a promising way to store distributed data and is gaining in popularity.

A distributed storage system for structured data called Bigtable is presented in [9]. The system is used for managing data that is designed to scale to very large size datasets distributed across thousands of commodity servers. Bigtable has successfully provided a flexible, high performance solution for all of the Google products. A significant amount of research has been done on database compression and distributed main memory database as described in [10-18].

2.1 Existing HIBASE Compression Technique

The basic HIBASE model, as described in [2], represents tables as a set of columns (rather than as a set of rows as used in a traditional relational database). This structure is dictionary based, and designed for low cardinality of domain values. The architecture places codes rather than the original data values, in tuples. The main pitfall of this method is that it cannot handle large databases because of its fully main memory dependency. HIBASE uses single block column vector; each attribute is associated with a domain dictionary and a column vector. The columns are organized as a linked list, each of which points into the dictionary. Fig. 1 shows a HIBASE structure together with its domain dictionary. There are 7 distinct lastnames represented by identifiers numbered 0 to 6 which can be represented by 3 bits; similarly suburb, state and marital status are represented by 2, 1 and 1 bits respectively. Hence, in compressed representation 7 bits are required to represent one tuple using the HIBASE method. For the set of 8 tuples, 56 bits would be required. In the uncompressed relation (Fig. 1) an average of 10 bits are required for each attribute, totalling 40 bits for each tuple, hence the total uncompressed relation requires 320 bits for all tuples.

The HIBASE approach thus appears to achieve a huge compression ratio; in fact the overall compression is somewhat less impressive because representing the domain dictionary requires some memory space.

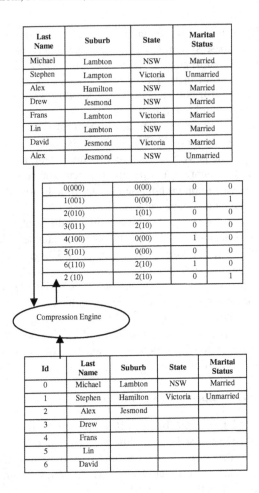

Last Name	Suburb	State	Marital Status
Michael	Lambton	NSW	Married
Stephen	Lampton	Victoria	Unmarried
Alex	Hamilton	NSW	Married
Drew	Jesmond	NSW	Married
Frans	Lambton	Victoria	Married
Lin	Lambton	NSW	Married
David	Jesmond	Victoria	Married
Alex	Jesmond	NSW	Unmarried

0(000)	0(00)	0	0
1(001)	0(00)	1	1
2(010)	1(01)	0	0
3(011)	2(10)	0	0
4(100)	0(00)	1	0
5(101)	0(00)	0	0
6(110)	2(10)	1	0
2 (10)	2(10)	0	1

Compression Engine

Id	Last Name	Suburb	State	Marital Status
0	Michael	Lambton	NSW	Married
1	Stephen	Hamilton	Victoria	Unmarried
2	Alex	Jesmond		
3	Drew			
4	Frans			
5	Lin			
6	David			

Fig. 1. Compressed relation with domain dictionary

3   Proposed Single Vector Large Data Cardinality (SVLDCS) Structure

The single vector large data cardinality structure (SVLDCS) is disk based, supports large data cardinality of domain dictionaries, and can used in a distributed environment. In this structure an Attribute_Bit_Storing dictionary is used to store the length of the required bit sequence for each attribute. The system generates a serial number for each tuple in the original relational table, and does not provide lineation of columns for different attributes as seen in the HIBASE method.

The (possibly huge amount of) information in a relational table is stored in compressed form with partitioning the attributes into different blocks. In each tuple of the compressed relation there may be a different number of blocks, and it is possible to store the information of a large number of attributes in each block. Each of the database fragments can accommodate 2^{32} tuples. Searching techniques can be applied to this compressed database format, and the actual information then retrieved from the domain dictionary. As the database gradually increases, the domain dictionaries can be partitioned so there is one (active) part in main memory and other (inactive) parts in permanent storage. A single vector is used to point to each of the database fragments.

When the number of tuples and attributes gradually increases, the single vector continues to be sufficient to handle this large data cardinality. The vector represents a collection of different fragments with multiple blocks and a large number of tuples. Any fragment of the compressed database can be distributed into any of the sites of

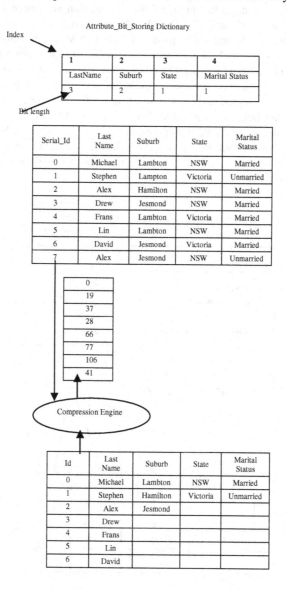

Fig. 2. Compressed Structure Using Single Block

the distributed environment, and a copy of the whole compressed database can be stored on disk.

Among the set of fragments there will at any time be a limited number of fragments in main memory. To satisfy query demand, other fragments may need to become available in main memory. New search key values (lexeme) are always inserted at the end of the domain dictionary. Encoding is performed before inserting the lexeme to make sure the lexeme is not redundant. The Ids of the search key values are retrieved from the original database relation before starting a query, after which the query results are found from the domain dictionary. In Fig. 1, all the information is compressed using binary values in the domain (columns) [2]. The SVLDCS approach represents this same information using a single columnar block as shown in Fig. 2. When the number of attributes and tuples increases, the SVLDCS structure is capable of representing this information using multiple blocks and compressed fragments are pointed to by a single vector.

The structure as used in a distributed environment is given in Fig. 3. The SVLDCS data structure is represented in Fig. 4. Each fragment can consist of up to 2^{32} tuples, and each block can hold up to 32 bits of information. While only 4 blocks are presented in Fig. 4, the same arrangement may be used to handle more blocks as well as a larger number of attributes. In Fig. 4, the single vector structure points to each compressed database fragment, each of which in turn is used to hold the information of a large number of tuples.

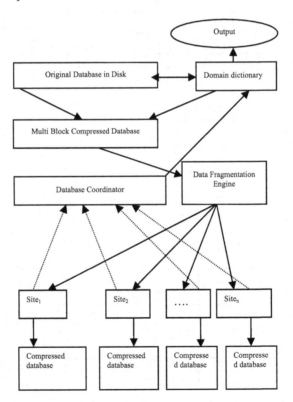

Fig. 3. Overall Structure of SVLDCS using Distributed Environment

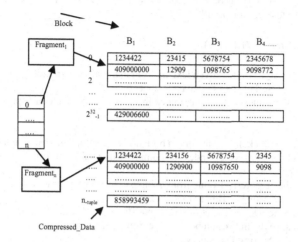

Fig. 4. Single vector structure with multi block compressed data representation

4   Searching Technique

The algorithm as described in Fig. 5 is used to handle large amounts of data using a single vector consisting of multiple fragments with multiple blocks. This algorithm, which can be used to search large data cardinality structure, is called SVLDC; *Single Vector Large Data Cardinality structure*. To understand the algorithm the following data structures are necessary:

Data Structure

AttributeBitStoring[]: Stores the required bit length for each attribute of the relation.

Lexeme: Value of the search key attribute.

Token: The Id of the encoded lexeme; retrieved lexeme: desired key (lexeme) values.

Domain_dictionary [ ] [ ]: Stores the distinct tuples for each attribute with token value.

Compressed_Data[ ]: Stores the results in compressed format.

Vector_index: Points to each fragment of the compressed database.

Single_vector [ ]: Stores the index of a vector up to n; where n= total number of tuples in main relation / 2^{32}.

Y[]: Array that stores all the compressed decimal values of specific tuple from a specific fragment.

X[]: Array used to store the binary representation of the decimal value that is stored in Y.

Single Column: Used to store the information of the relation when the number of tuples and attributes are not on a large scale.

Algorithm Search_SVLDCS()
Begin
  Search the serial_Id for the given lexeme from main database relation;
  Vector_Index= Serial_Id of the given lexeme/ 2^{32};
  //Accessing the values of multiple //block from the specific tuple of //specific fragment
  For (i=0 to p-1 Block do)
    Begin
  // where p (total blocks)=   //maximum bit length of the tuple //of main relation/32;
    Y[i]=Single_vector[vector_index]-> compressed_data[serial_Id][i];
    End;
  //retrieving the compressed value from //single columnar vector table;
  X= X_1X_2.......X_n=Int_to_Binary [Yi .....Yp];
  Retrieved_lexeme=Domain_dictionary[source][tagret];
    Where **target**= the index number of target (output)attribute from
    Attribute_Bit_Storing Dictionary;
    first_bit_length= Σ Bit length (from starting bit length to length of query attribute from
  Attribute_Bit_Storing Dictionary);
    source=Converted decimal value of X[],from position of the first_bit_length to number
  of bit length of query attribute;
    If(retrieved_lexeme) return 1; //Value found
      Else return 0;
    // or search for another input;
  End. // End of Main.

Fig. 5. Algorithm Single Vector Structure for large compressed database

4.1 Explanation of the Searching Technique of (SVLDCS) Structure

Consider the original relation given in Fig. 1, Attribute_Bit_Storing Dictionary, Domain_dictionary, and Single_Column as given in Fig. 2, suppose it is required to find the State information of LastName= Drew. The system finds the Serial Id of the lexeme (LastName from original relation), which is 3. The value of that index (Serial Id) position from Single_Column is 28.

So Y=Single_Column [3] = 28. The converted binary value of Y is stored in X. Thus X=X_0.....X_6=Int_to_Binary (Y) =0011100 (as the length bit is 7).

Retrieved_lexeme (Value of State) = Domain_Dictionary [source] [target], where target = the index number of the (State) attribute from Attribute_Bit_Storing_Dictionary=3

source=Converted decimal value of X [ ], from position of the first_bit_length to the number of bit length of retrieved_lexeme.

where first_bit_length= Σ Bit length (From the starting bit length upto the length of retrieved lexeme (from Attribute_Bit_Storing Dictionary)).

Firstbitlength=3+2=5 and length of the attribute state is 1.

Source= Converted decimal value of X [ ] [From the 5th position to 5th position, as length of attribute state is 1)=Bin_to_Decimal (0)=0

Retrieved_lexeme (Value of State) = Domain Dictionary [source] [target] = Domain_Dictionary [0] [3]= NSW. We see the value of the retrieved lexeme (State)

=NSW which is also the same in the original database relation. Similarly this technique is applied to the large number of tuples with multiple fragments and attributes divided into blocks according to the given algorithm in Fig. 5. In Fig. 4, the information for a large number of attributes and tuples are presented, providing multiple fragments and blocks with a single vector.

4.2   Searching Time Analysis of SVLDC Structure

The total search time of SVLDC structure is (T_{SVLD}) = (Time taken to search the lexeme Id from original relation + Domain Dictionary searching + Compressed_Data searching from Single vector):

$$T_{SVLD} = T_{LOR} + T_{DD} + T_{CD} \tag{1}$$

In a compressed_data search from a single vector, a hashing technique is applied to find the vector index as well as fragment location (given as fragment_no=hash (serial_id_lexeme)/2^{32}). So the compressed_data searching time is:

$$T_{CD} = O(1) \tag{2}$$

In the domain dictionary only the value of the particular attributes are retrieved during a search of the structure. Therefore the search time of the domain dictionary is constant. To insert a new tuple in the database, the domain dictionary would be searched to make sure of its existence in the database; if the lexeme is not found in the domain dictionary, the new lexeme is inserted at the end of the domain dictionary and a token is created for that lexeme. So

$$T_D = O(1) + O(n) \tag{3}$$

where n is the total number of tuples in the domain dictionary. To find out the Serial_Id of the lexeme from the main relation the required time would be:

$$T_{LOR} = O(n * \log_2^n + \log_2^n) \tag{4}$$

A binary search technique may be applied to find any lexeme's serial Id and it would take $O(\log_2^n)$ time, where n is the total number of tuples. Before applying a binary search technique, it is required to sort the relation according to any specific attribute, and that takes O ($n * \log_2^n$) time.

5   The Analysis of Storage Capacity of SVLDC Structure

The required memory for each Compressed Data fragment is:

$$S_{CF} = \sum (m * p * b) \tag{5}$$

where $m=$ maximum_no_of tuples of a fragment$=2^{32}$; p=no_of_block_per_tuple; b=avg_bytes_required_each_block; n= Total no of tuples in main relation /m; Fragment_tuple= Total no of tuple in main relation % m;

When (Fragment_tuple) =0, this represents that every fragment in a single vector is filled up with the maximum number of tuples.

Hence the required memory for a Single Vector is:

$$S_{SV} = \sum_{j=1}^{n} (S_{CF}) \tag{6}$$

The required memory for the Domain Dictionary is:

$$S_{DD} = \sum_{i=1}^{q} (C * L) \tag{7}$$

where q is the total number of tuples in domain dictionary, C is the number of attributes, L is the average length of each attribute. Combining equations (6) and (7) the total required memory for SVLDCS is:

$$S_{SVLDC} = S_{SV} + S_{DD} \tag{8}$$

When (Fragment_tuple) is not zero, this indicates that all the fragments are not filled by the maximum number of tuples. It is convention that the last fragment has tuples that are less than the maximum number of tuples. In that case the required memory is:

$$S_{SVLDC} = [\sum_{j=1}^{n} (S_{CF}) + S_{DD} + S_{NF}] \tag{9}$$

where S_{NF} =fragment_tuple*no_of_blocks_per_tuple*bytes_required_for_each_block.

5.1 Analytical Analysis of Storage Capacity Using Different Methods

Let CF be the compression factor, $UR_{Storage}$ be the required storage for uncompressed relation, $CRV_{Storage}$ be the storage capacity for compressed relation in SVLDCS, S_{DD} be the domain dictionary storage capacity. The compression factor is represented by

$$CF = UR_{storage} / (CRV_{storage} + S_{DD}) \tag{10}$$

It is estimated that if the dictionary takes 25% of total storage, then $S_{DD} = .25 UR_{Storage}$; $CRV_{Storage} = .75 UR_{Storage}$;

So, $S_{DD} / CRV_{Storage} = .25/.75 =.33$;

$$S_{DD} =.33 * CRV_{Storage} \tag{11}$$

Combining (10) and (11) we have

$$CF = UR_{storage} \big/ 1.33 * CRV_{storage}$$

$$UR_{Storage} = CF * 1.33 * CRV_{storage} \tag{12}$$

Oracle data compression [7] achieves the average compression factor $CF_{orc} = 3.11$. The total storage required in the Oracle compression technique: $ORC_{Storage}$:

= the required storage for uncompressed data / Compression factor

$$ORC_{Storage} = UR_{Storage} / CF_{orc;}$$

Using equation (12) we have:

$$ORC_{Storage} = CF * 1.33 * CRV_{storage} / 3.11$$

$$ORC_{Storage} = CF * .43 * CRV_{storage} \tag{13}$$

In [19] the Tera-scale architecture is described on a Dual processor, achieving a data rate in compressed form of 3GB/day, whereas uncompressed data stream is about 15GB/day, so compression (CF_{tera}) factor is about 5. So the storage capacity for Tera-scale is:

$$Tera_{storage} = UR_{Storage} / CF_{tera} ;$$

Using equation (12) we have

$$Tera_{storage} = CF * 1.33 * CRV_{storage} / 5 ;$$

$$Tera_{storage} = 1.596 * CRV_{Storage} \tag{14}$$

Table 1. Required Memory (in Terabytes) using Different Methods

No_of Tuples million	Uncomp ressed (TB)	ORACLE (TB)	TERA-SCALE (TB)	SVLDCS (TB)
759	0.1875	0.059	0.037	0.023
1215	0.3	0.096	0.06	0.037 5
1620	0.4	0.129	0.08	0.05
2025	0.5	0.161	0.1	0.062 6
3038	0.75	0.239	0.15	0.093
3443	0.85	0.273	0.17	0.106
3848	0.95	0.307	0.19	0.119
4172	1.03	0.335	0.207	0.13
5388	1.33	0.43	0.266	0.167
8600	2. 20	0.71	0.44	0.275

Graphical representations of memory requirements of uncompressed database, SVLDCS, compressed database in Oracle [7] and compressed database in Tera-scale [19] structure are presented in Fig. 6. It is clear that the storage size of databases increases due to increase in the number of tuples. The number of tuples (in millions) are represented by the X axis and the storage capacity (in Tera Bytes) of different methods are represented by the Y axis. The storage comparison is also presented in a tabular form (in Table 1) for different methods using equation (8), (9), (12), (13) and (14). Note that Oracle [7] and Tera-Scale ([19] requires 710 GB and 440 GB respectively while SVLDCS reduces these significantly to only 275 GB. From Table 1, we see the compression factor is almost 8:1, comparing uncompressed to compressed relation (SVLDCS). After considering the domain dictionary (using equation (12)), this compression factor becomes 6:1.

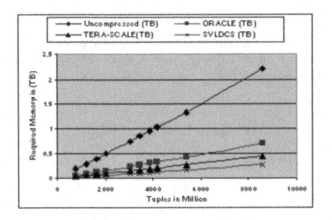

Fig. 6. Comparison of Storage Space for Different Methods

6 Conclusions

Management and processing of large scale data sets is time-consuming, costly and an obstruction to research. Accessing a rapidly evolving large scale database in a concurrent environment is also challenging because the number of disk accesses increases as the database size grows; the response time of any query also increases. This paper describes an innovative disk based single vector large data cardinality approach, incorporating data compression in a distributed environment. According to the technique data are stored in a more space-efficient way than is provided by other existing schemes such as HIBASE, Oracle and Tera-scale. The compressed structure of SVLDCS is better than compression in Oracle, because Oracle applies block level compression that includes some redundancy. The Tera-scale architecture also compress data file but querying is not possible when the data is in compressed form. When the SVLDC approach is applied to a database of 2.2 TB (tera bytes), only 275 GB of storage is required, a significant improvement over other schemes. The compression factor achieved using the SVLDCS structure is 6:1 compared to uncompressed relation.

Querying, updating, inserting, deleting, and searching data in the databases is supported by the SVLDCS technique; details of these will be further reported as research progresses.

References

1. Garcia-Molina, H., Salem, K.: Main Memory Database Systems: An Overview. IEEE Transaction on Knowledge and Data Engineering 4(6), 509–516 (1992)
2. Cockshott, W.P., Mcgregor, D., Wilson, J.: High-Performance Operations Using a Compressed Database Architecture. The Computer Journal 41(5), 283–296 (1998)
3. Pucheral, P., Thevnin, J.-M., Valduriez, P.: Efficient Main Memory Data Management using DBGraph Storage Model. In: The 16th International Conference on Very Large Databases, Brisbase, Australia (1990)
4. Hoque, A.S.M.L.: Storage and Querying of High Dimensional Sparsely Populated Data in Compressed Representation. In: Shafazand, H., Tjoa, A.M. (eds.) EurAsia-ICT 2002. LNCS, vol. 2510, p. 418. Springer, Heidelberg (2002)
5. Alkhatib, G., Labban, R.S.: Transaction Management in Distributed Database Systems: the Case of Oracle's Two-Phase Commit. The Journal of Information Systems Education 13(2), 95–103 (1995)
6. Lawrence, R., Kruger, A.: An Architecture for Real-T'ime Warehousing of Scientific Data. In: The International Conference on Scientific Computing (ICSC), Vegus, Nevada (2005)
7. Poess, M., Potapov, D.: Data Compression in Oracle. In: The 29th International Conference on Very Large Databases(VLDB), Berlin, Germany (2003)
8. Litwin, W., Moussa, R., Thomas, J.E., Schwartz, S.J.: LH*RS: A Highly Available Distributed Data Storage. In: The 30th International Conference on Very Large Databases Conference, Toronto, Canada (2004)
9. Chang, F., Dean, J., Ghemawat, S., Hsieh, W.C., Wallach, D.A., Burrows, M., Chandra, T., Fikes, A., Gruber, R.E.: Bigtable: A Distributed Storage System for Structured Data. In: The International Conference on Operating Systems Design and Implementation (OSDI), Seattle, Wa, USA (2006)
10. Hoque, A.S.M.l., McGregor, D., Wilson, J.: Database compression using an off-line dictionary method. In: Yakhno, T. (ed.) ADVIS 2002. LNCS, vol. 2457. Springer, Heidelberg (2002)
11. Lee, I., Yeom, H.Y., Park, T.: A New Approach for Distributed Main Memory Database Systems: A Casual Commit Protocol. IEICE Trans. Inf. & Syst. 87(1), 196–204 (2004)
12. Lehman, T.J., Shekita, E.J., Cabrera, L.-F.: An Evaluation of Starburst's Memory Resident Storage Component. IEEE Transaction on Knowledge and Data Engineering, 555–566 (1992)
13. Lim, H.-S., Lee, J.-G., Lee, M.-J., Whang, K.-Y., Song, I.-Y.: Continuous Query Processing in Data Streams Using Duality of Data and Queries. In: SIGMOD Chicago, Illinois, USA (2006)
14. Liu, F., Yu, C., Meng, W., Chowdhury, A.: Effective Keyword Search in Relational Databases SIGMOD Chicago, Illinois, USA (2006)
15. Liu, X., Li, X.: Design and Implement of Distributed Database-based Pricing Management System*. In: Proceedings of the 6th World Congress on Intelligent Control and Automation, Dalian, China (2006)

16. Pucheral, P., Thevenin, J.-M., Valduriez, P.: Efficient Main Memory Data Management using DBGraph Storage Model. In: The 16th International Conference on Very Large Databases(VLDB), Brisbase, Australia (1990)
17. Teorey, T.J.: Distributed Database Design: A Practical Approach and Example. SIGMOD 18(4), 23–39 (1989)
18. Valduriez, P., Ozsu, T.: Principle of Distributed Database Systems. Prentice Hall, Englewood Cliffs (1999)
19. Lawrence, R., Kruger, A.: An Architecture for Real-Time Warehousing of Scientific Data. In: The International Conference on Scientific Computing (ICSC), Vegus, Nevada, USA (2005)

Relaxed Approaches for Correct DB-Replication with SI Replicas*

J.E. Armendáriz-Iñigo[1], J.R. Juárez-Rodríguez[1], J.R. González de Mendívil[1],
J.R. Garitagoitia[1], F.D. Muñoz-Escoí[2], and L. Irún-Briz[2]

[1] Universidad Pública de Navarra, 31006 Pamplona, Spain
{enrique.armendariz,jr.juarez,mendivil,joserra}@unavarra.es
[2] Instituto Tecnológico de Informática, 46022 Valencia, Spain
{fmunyoz,lirun}@iti.upv.es

Abstract. The concept of Generalized Snapshot Isolation (GSI) has been recently proposed as a suitable extension of conventional Snapshot Isolation (SI) for replicated databases. In GSI, transactions may use older snapshots instead of the latest snapshot required in SI, being able to provide better performance without significantly increasing the abortion rate when write/write conflicts among transactions are low. We study and formally proof a sufficient condition that replication protocols with SI replicas following the deferred update technique must obey to achieve GSI. They must provide global atomicity and commit update transactions in the very same order at all sites. However, as this is a sufficient condition, it is possible to obtain GSI by relaxing certain assumptions about the commit ordering of certain update transactions.

1 Introduction

Snapshot Isolation (SI) is the isolation level provided by several commercial database systems, such as Oracle and PostgreSQL. Transactions executed under SI allow to read from the last committed snapshot and, hence, read operations are never blocked nor conflict with any other update transaction. In order to prevent the lost update phenomenon [1], concurrent update transactions (read-only transactions are always committed) modifying the same data item apply the *first-committer-wins* rule: only the first transaction that commits is allowed to proceed the remainder are aborted. This turns out into a nice feature because it provides sufficient data consistency (though not serializable [2,3]) for non-critical applications while it maintains a good performance, since read-only transactions are neither delayed, blocked nor aborted and they never cause update transactions to block or abort. This behavior is important for workloads dominated by read-only transactions, such as those resulting from dynamic content Web servers [4].

Many enterprise applications demand high availability since they have to provide continuous service to their users. This also implies to replicate the information being used; i.e., to manage replicated databases. The concept of Generalized Snapshot Isolation (GSI, concurrently to this a similar definition denoted as 1-copy-SI was proposed

---

* This work has been supported by the EU FEDER and Spanish MEC under research grant TIN2006-14738-C02.

J. Cordeiro et al. (Eds.): ICSOFT 2008, CCIS 47, pp. 161–174, 2009.

in [5]) has been recently proposed [3] in order to provide a suitable extension of conventional SI for replicated databases based on multiversion concurrency control. In GSI, transactions may use older snapshots instead of the latest snapshot required in SI (setting up the latest snapshot in a distributed setting is not trivial). Actually, authors of [3] outline an impossibility result which justifies the use of GSI in database replication: "*there is no non-blocking implementation of SI in an asynchronous system, even if databases never fail*" which has been formally justified in [6].

The deferred update technique [7] consists in executing transactions at their delegate replicas (obtaining their corresponding snapshot) and setting up a commit ordering for update transactions which is mainly done thanks to the total order broadcast [8]. When a transaction requests its commitment (read-only transactions are committed right away) its updates are collected and broadcast (using the total order primitive) to the rest of replicas. Upon its delivery at replicas a validation test (i.e. to detect conflicts with other concurrent transactions in the system) is performed; namely a certification test [9] that performs the distributed first-committer-wins rule [3,5] in the same way at al replicas and ensures the same order of the commit process of transactions. The main advantage of these replication protocols is that transactions can start at any time without restriction or delay.

In this paper, we formalize the requirements for achieving GSI over SI replicas using *non-blocking* protocols. Thus, the criteria for implementing GSI are: (i) Each submitted transaction to the system either commits or aborts at all sites (*atomicity*); (ii) All update transactions are committed in the same total order at every site (*total order of committed transactions*). Total order ensures that all replicas see the same sequence of transactions, being thus able to provide the same snapshots to transactions, independently of their starting replica; i.e. giving the logical vision of a one copy scheduler (1-Copy-GSI). Whereas atomicity guarantees that all replicas take the same actions regarding each transaction, so their states should be consistent, once each transaction has been terminated.

One can think that these assumptions are rather intuitive but they constitute the milestone for our contribution of the paper. It consists in somehow relaxing the assumption of the total order of committed transactions. If a protocol is not careful about that, those transactions without write/write conflicts might be applied in different orders in different replicas. So, transactions would be able to read different versions in different replicas. However, this optimization is important since processing messages serially as supposed for replication protocols deployed over a group communication system [8] would result in significantly lower throughput rates. A relaxing assumption has been already presented in [5], still using the total order broadcast, it lets validated transactions to apply (and commit) transactions concurrently as long as their respective updates do not intersect. However, this protocol needs to *block* the execution of the first operation of any starting transaction until the concurrent application of transactions finishes. Thus, it is easy to see that there are multiple approaches to obtain GSI at the price of imposing certain restrictions, in particular, the need to block the start of transactions to obtain a global consistent snapshot. Finally, we take a look and discuss how to relax this last contribution, which is actually too strong, for deploying GSI non-blocking protocols.

The rest of the work is organized as follows[1]. Section 2 introduces the concept of multiversion histories based on [10]. Section 3 gives the concept of GSI. In Section 4, the structure of deferred update replication protocols is introduced. Conditions for 1-Copy-GSI is introduced in 5. We take a look at how to relax conditions for 1-Copy-GSI in Section 6. Finally, conclusions end the paper.

2 Multiversion Histories

In the following, we define the concept of multiversion history for committed transactions using the theory provided in [10]. The properties studied in our paper only require to deal with committed transactions. To this end, we first define the basic building blocks for our formalizations, and then the different definitions and properties will be shown.

A database (DB) is a collection of data items, which may be concurrently accessed by transactions. A history represents an *overall partial ordering* of the different operations concurrently executed within the *context* of their corresponding transactions. Thus, a multiversion history generalizes a history where the database items are versioned.

To formalize this definition, each transaction submitted to the system is denoted by T_i. A transaction is a sequence of read and write operations on database items ended by a commit or abort operation. Each T_i's write operation on item X is denoted $W_i(X_i)$. A read operation on item X is denoted $R_i(X_j)$ stating that T_i reads the version of X installed by T_j. Finally, C_i and A_i denote the T_i's commit and abort operation respectively. We assume that a transaction does not read an item X after it has written it, and each item is read and written at most once. Avoiding redundant operations simplifies the presentation. The results for this kind of transactions are seamlessly extensible to more general models. In any case, redundant operations can be removed using local variables in the program of the transaction [11].

Each version of a data item X contained in the database is denoted by X_i, where the subscript stands for the transaction identifier that installed that version in the DB. The *readset* and *writeset* (denoted by RS_i and WS_i respectively) express the sets of items read (written) by a transaction T_i. Thus, T_i is a *read-only* transaction if $WS_i = \emptyset$ and it is an *update* one, otherwise.

Let $T = \{T_0, T_1, \ldots, T_n\}$ be a set of *committed* transactions, where the operations of T_i are ordered by $\prec_{T_i}$. The last operation of a transaction is the commit operation. We assume that transaction T_0 is an initial transaction. T_0 is executed and committed before any other transaction and initializes all data items $X \in DB$ to X_0. To process operations from a transaction $T_i \in T$, a multiversion scheduler must translate T_i's operations on data items into operations on specific versions of those data items. That is, there is a function h that maps each $W_i(X)$ into $W_i(X_i)$, and each $R_i(X)$ into $R_i(X_j)$ for some $T_j \in T$.

Definition 1. *A Complete Committed Multiversion (CCMV) history H over T is a partial order with order relation $\prec$ such that:*

---

[1] Due to space constraints, the reader is referred to [6] for a thorough explanation of the correctness proof.

(1) $H = h(\bigcup_{T_i \in T} T_i)$ *for some translation function* h.
(2) $\prec \supseteq \bigcup_{T_i \in T} \prec_{T_i}$.
(3) If $R_i(X_j) \in H$, $i \neq j$, *then* $W_j(X_j) \in H$ *and* $C_j \prec R_i(X_j)$.

In the previous Definition 1 condition (1) indicates that each operation submitted by a transaction is mapped into an appropriate multiversion operation. Condition (2) states that the CCMV history preserves all orderings stipulated by transactions. Condition (3) establishes that if a transaction reads a concrete version of a data item, it was written by a transaction that committed before the item was read.

Definition 1 is more specific than the one stated in [10], since the former only includes committed transactions and explicitly indicates that a new version may not be read until the transaction that installed the new version has committed. In the rest of the paper, we use the following conventions: (i) $T = \{T_0, T_1, \ldots, T_n\}$ is the set of committed transactions for every defined history; and, (ii) any history H is a CCMV history over T. Note that these conventions will be also applicable when a superscript is used to denote the site of the database where the history is generated.

In general, two histories $(H, \prec)$ and $(H', \prec')$ over the same set of transactions are *view equivalent* [10], denoted as $H \equiv H'$ if they contain the same operations, have the same *reads-from* relations, and produce the same final writes. The notion of equivalence of CCMV histories reduces to the simple condition, $H = H'$, if the following *reads-from* relation is used: T_i *reads* X *from* T_j in a CCMV history $(H, \prec)$, if and only if $R_i(X_j) \in H$.

3 Generalized Snapshot Isolation

In SI reading from a snapshot means that a transaction T_i sees all the updates done by transactions that committed before the transaction started its first operation. The results of its writes are installed when the transaction commits. However, a transaction T_i will successfully commit if and only if there is not a concurrent transaction T_k that has already committed and some of the written items by T_k are also written by T_i. From our point of view, histories generated by a given concurrency control providing SI may be interpreted as multiversion histories with time restrictions.

On the other hand, the concept of Generalized Snapshot Isolation (GSI) was firstly applied to database replication in [3]. A hypothetical concurrency control algorithm could have stored some past snapshots. A transaction may receive a snapshot that happened in the system before the time of its first operation (instead of its current snapshot as in a SI concurrency control algorithm). The algorithm may commit the transaction if no other transaction impacts with it from that past snapshot. Thus, a transaction can observe an older snapshot of the DB but the write operations of the transaction are still valid update operations for the DB at commit time.

Definition 2. *Let* $(H, \prec)$ *be a history and* $t \colon H \to \mathbb{R}^+$ *a mapping such that it assigns to each operation* $op \in H$ *its real time occurrence* $t(op) \in \mathbb{R}^+$. *The schedule* H_t *of the history* $(H, \prec)$ *verifies:*
(1) If $op, op' \in H$ *and* $op \prec op'$ *then* $t(op) < t(op')$.
(2) If $t(op) = t(op')$ *and* $op, op' \in H$ *then* $op = op'$.

The mapping $t()$ totally orders all operations of $(H, \prec)$. Condition (1) states that the total order $<$ is compatible with the partial order $\prec$. Condition (2) establishes, for sake of simplicity, the assumption that different operations will have different times. We are interested in operating with schedules since it facilitates the work, but only with the ones that derive from CCMV histories over a concrete set of transactions T. One can note that an arbitrary time labeled sequence of versioned operations, e.g. $(R_i(X_j), t_1), (W_i(X_k), t_2)$ and so on, is not necessarily a schedule of a history. Thus, we need to put some restrictions to make sure that we work really with schedules corresponding to possible histories.

Property 1. *Let S_t be a time labeled sequence of versioned operations over a set of transactions T, S_t is a schedule of a history over T if and only if it verifies the following conditions:*
(1) there exists a mapping h such that $S = h(\bigcup_{i \in T_i} T_i)$.
(2) if $op, op' \in T_i$ and $op \prec_{T_i} op'$ then $t(op) < t(op')$ in S_t.
(3) if $R_i(X_j) \in S$ and $i \neq j$ then $W_j(X_j) \in S$ and $t(C_j) < t(R_i(X_j))$.
(4) if $t(op) = t(op')$ and $op, op' \in S$ then $op = op'$.

The proof of this fact can be inferred trivially. In the following, we use an additional convention: (iii) A schedule H_t is a schedule of a history $(H, \prec)$. Note that every schedule H_t may be represented by writing the operations in the total order $(<)$ induced by $t()$. We define the "commit time" (c_i) and "begin time" (b_i) for each transaction $T_i \in T$ in a schedule H_t as $c_i = t(C_i)$ and $b_i = t(\textit{first operation of } T_i)$, holding $b_i < c_i$ by definition of $t()$ and $\prec_{T_i}$. In the following, we formalize the concept of snapshot of the database. Intuitively, it comprises the latest version of each data item. Firstly, we will see an example of this:

Example 1. *Let us consider the following transactions T_1, T_2 and T_3: $T_1 = R_1(X) W_1(X) c_1$, $T_2 = R_2(Z) R_2(X) W_2(Y) c_2$, $T_3 = R_3(Y) W_3(X) c_3$. A sample of a possible schedule of these transactions might be the following one: $b_1 R_1(X_0) W_1(X_1) c_1 \ b_2 R_2(Z_0) b_3 R_3(Y_0) W_3(X_3)$ $c_3 R_2(X_1) W_2(Y_2) c_2$. As this example shows, each transaction is able to include in its snapshot (and read from it) the latest committed version of each existing item at the time such transaction was started. Thus T_2 has read version 1 of item X since T_1 has generated such version and it has already committed when T_2 started. But it only reads version 0 of item Z since no update of such item is seen by T_2. This is true despite transactions T_2 and T_3 are concurrent and T_3 updates X before T_2 reads such item, because the snapshot taken for T_2 is previous to the commit of T_3.*

This example provides the basis for defining what a snapshot is. For that purpose, we need to define first the set of installed versions of a data item X in a schedule H_t, as the set $Ver(X, H) = \{X_j : W_j(X_j) \in H\} \cup X_0$, being X_0 its initial version.

Definition 3. *The snapshot of the database DB at time $\tau \in \mathbb{R}^+$ for a schedule H_t, is defined as: $Snapshot(DB, H_t, \tau) = \bigcup_{X \in DB} latestVer(X, H_t, \tau)$ where the latest version of each item $X \in DB$ at time τ is the set: $latestVer(X, H_t, \tau) = \{X_p \in Ver(X, H) : (\nexists X_k \in Ver(X, H) : c_p < c_k \leq \tau)\}$*

From the previous definition, it is easy to show that a snapshot is modified each time an update transaction commits. If $\tau = c_m$ and $X_m \in Ver(X, H)$, then $latestVer(X, H_t,$

$c_m) = \{X_m\}$. In order to formalize the concept of GSI-schedule, we utilize a slight variation of the predicate *impacts* for update transactions presented in [3]. Two transactions $T_j, T_i \in T$ impact at time $\tau \in \mathbb{R}^+$ in a schedule H_t, denoted T_j *impacts* T_i at τ, if the following predicate holds: $WS_j \cap WS_i \neq \emptyset \wedge \tau < c_j < c_i$.

Definition 4. *A schedule H_t is a GSI-schedule of the database DB if and only if for each $T_i \in T$ there exists a value $s_i \in \mathbb{R}^+$ such that $s_i \leq b_i$ and:*
(1) if $R_i(X_j) \in H$ then $X_j \in Snapshot(DB, H_t, s_i)$; and,
(2) for each $T_j \in T$: $\neg(T_j$ impacts T_i at $s_i)$.

Condition (1) states that every item read by a transaction belongs to the same (possible past) snapshot. Condition (2) also establishes that the time intervals $[s_i, c_i]$ and $[s_j, c_j]$ do not overlap for any pair of write/write conflicting transactions T_i and T_j. If for all $T_i \in T$, conditions (1) and (2) hold for $s_i = b_i$ then H_t is a SI-schedule. Another observation of the definition concludes that if there exists a transaction $T_i \in T$ such that conditions (1) and (2) are only verified for a value $s_i < b_i$ then there is an item $X \in RS_i$ for which $latestVer(X, H_t, s_i) \neq latestVer(X, H_t, b_i)$. That is, the transaction T_i has not seen the latest version of X at the begin time b_i. There was a transaction T_k with $W_k(X_k) \in H$ such that $s_i < c_k < b_i$. This can be best seen in the next example.

Example 2. *The following is an example of a GSI-schedule: $b_1 R_1(X_0) W_1(X_1) c_1 b_2 R_2(X_0)$ $R_2(Z_0) b_3 R_3(Y_0) W_3(X_3) c_3 W_2(Y_2) c_2$. In this schedule, transaction T_2 reads X_0 after the commit of T_1 appears. This would not be correct for a SI-schedule (since the read version of X is not the latest one), but it is perfectly valid for a GSI-schedule, taken the time point of the snapshot provided to T_2 (i.e. s_2) previous to the commit of T_1, as it is shown: $b_1 R_1(X_0)$ $\mathbf{s_2}$ $W_1(X_1) c_1 b_2 R_2(X_0) R_2(Z_0) b_3 R_3(Y_0) W_3(X_3) c_3 W_2(Y_2) c_2$. The intuition under this schedule in a distributed system is that the message containing the modifications of T_1 (the write operation on X) would have not yet arrived to the site at the time transaction T_2 began. This may be the reason for T_2 to see this previous version of item X. The fact that GSI captures these delays into schedules makes attractive its usage on distributed environments.*

It is possible to think that if the operations in the GSI-schedule obtained from the history H had been "on time" then the schedule would have been a SI-schedule. The next property states that the aforementioned transformation does always exist.

Property 2. *Let H_t be a GSI-schedule. There is a mapping $t': H \rightarrow \mathbb{R}^+$ such that $H_{t'}$ is a SI-schedule.*

This last property states that if H_t is a GSI-schedule, there will exist a $H_{t'}$, which is actually a SI-schedule, and verify the following $H_t \equiv H_{t'}$ (in the sense of view-equivalence).

4   The Deferred Update Technique

As it was mentioned in the previous section, GSI is particularly interesting in replicated databases, since many replication protocols execute each transaction initially in a delegate replica, propagating later its updates to the rest of replicas [5,3,12]. This means that transaction writesets cannot be immediately applied in all replicas at a time and, due to this, the snapshot being used in a transaction might be "previous" to the one that (regarding physical time in a hypothetical centralized system) would have been assigned to it. In this Section we consider a distributed system that consists of m sites, being $I_m = \{1..m\}$ the set of site identifiers. Sites communicate among them by reliable message passing. We make no assumptions about the time it takes for sites to execute and for messages to be transmitted. We assume a system free of failures. Each site k runs an instance of the database management system and maintains a copy of the database DB. We will assume that each database copy, denoted DB^k with $k \in I_m$, provides SI [1]. We use the transaction model of Section 2. Let $T = \{T_i : i \in I_n\}$ be the set of transactions submitted to the system; where $I_n = \{1..n\}$ is the set of transaction identifiers.

The deferred update technique defines for each transaction $T_i \in T$, the set of transactions $\{T_i^k : k \in I_m\}$ in which there is only one, denoted $T_i^{site(i)}$, verifying $RS_i^{site(i)} = RS_i$ and $WS_i^{site(i)} = WS_i$; for the rest of the transactions, T_i^k, $k \neq site(i)$, $RS_i^k = \emptyset$ and $WS_i^k = WS_i$. $T_i^{site(i)}$ determines the local transaction of T_i, i.e., the transaction executed at its delegate replica or site, whilst $T_i^k, k \neq site(i)$, is a remote transaction of T_i, i.e., the updates of the transaction executed at a remote site. An update transaction reads at one site and writes at every site, while a read-only transaction only exists at its local site. In the rest of the paper, we consider the general case of update transactions with non-empty sets.

Let $T^k = \{T_i^k : i \in I_n\}$ be the set of transactions submitted at each site $k \in I_m$ for the set T. Some of these transactions are local at k while others are remote ones. In the next, the Assumption 1 implies that each transaction submitted to the system either commits at all replicas or in none of them. Thus, the updates applied in a delegate replica by a given transaction are also applied in the rest of replicas. Obviously, we consider a fully-replicated system. Since only committed transactions are relevant, the histories being generated at each site should be histories over T^k, as defined above.

Assumption 1 (Atomicity). *H^k is a CCMV history over T^k for all sites $k \in I_m$.*

In the considered distributed system there is not a common clock or a similar synchronization mechanism. However, we can use a logical real time mapping $t : \bigcup_{k \in I_m}(H^k) \rightarrow \mathbb{R}^+$ that totally orders all operations of the system. This mapping is compatible with each partial order $\prec^k$ defined for H^k for each site $k \in I_m$. In the following, we consider that each DB^k provides SI-schedules under the previous time mapping.

Assumption 2 (SI Replicas). *H_t^k is a SI-schedule of the history H^k for all sites $k \in I_m$.*

In order to study the level of consistency implemented by this kind of non-blocking protocols it is necessary to define the one copy schedule (1C-schedule) obtained from the schedules at each site. In the next definitions, properties and theorems we use the following notation: for each transaction T_i, $i \in I_n$, $C_i^{min(i)}$ denotes the commit operation of the transaction T_i at site $min(i) \in I_m$ such that $c_i^{min(i)} = \min_{k \in I_m}\{c_i^k\}$ under the considered mapping $t()$.

Definition 5 (1C-schedule). *Let $T = \{T_i : i \in I_n\}$ be the set of submitted transactions to a replicated database system with a non-blocking deferred update strategy that verifies Assumption 1 and Assumption 2. Let $S = \bigcup_{k \in I_m}(H^k)$ be the set formed by the*

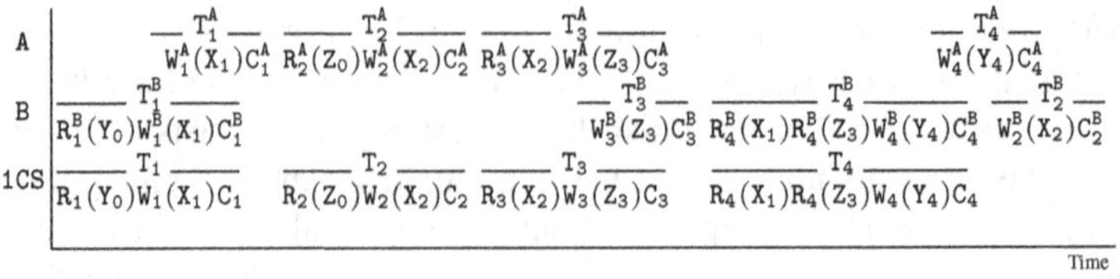

Fig. 1. Replicated one-copy execution not providing CSI nor GSI

union of the histories H^k over $T^k = \{T_i^k : i \in I_n\}$. And let $t: S \to \mathbb{R}^+$ be the mapping that totally orders the operations in S. The 1C-schedule, $H_{t'} = (H, t': H \to \mathbb{R}^+)$, is built from S, $t()$ and for each $i \in I_n$ and $k \in I_m$ as follows:
(1) Remove from S operations $W_i(X_i)^k$, with $k \neq site(i)$, and C_i^k, with $k \neq min(i)$.
(2) H is obtained with the rest of operations in S after step (1), applying the renaming:
$W_i(X_i) = W_i(X_i)^{site(i)}$; $R_i(X_j) = R_i(X_j)^{site(i)}$; and, $C_i = C_i^{min(i)}$.
(3) Finally, $t'()$ is obtained from $t()$ as follows: $t'(W_i(X_i)) = t(W_i(X_i)^{site(i)})$; $t'(R_i(X_j)) = t(R_i(X_j)^{site(i)})$; and, $t'(C_i) = t(C_i^{min(i)})$.

As $t'()$ receives its values from $t()$, we write, H_t instead of $H_{t'}$. In the 1C-schedule H_t, for each transaction T_i, is trivially verified $b_i < c_i$ because this technique guarantees that for all $k \neq site(i)$, $b_i^{site(i)} < b_i^k$. The 1C history H, that is formed by the operations over the logical DB, is also a history over T. We prove this fact informally. By the renaming (2) in Definition 5, each transaction T_i, has its operations over the data items in RS_i and WS_i, and $\prec_{T_i}$ is trivially maintained in a partial order $\prec$ for H, because H_t contains the local operations of $T_i^{site(i)}$. H is also formed by committed transactions, under Assumption 1; for each T_i, $C_i \in H$. Finally, if $R_i(X_j) \in H$, then $R_i(X_j)^{site(i)} \in H^{site(i)}$. As $H^{site(i)}$ is a history over $T^{site(i)}$ then $C_j^{site(i)} \prec R_i(X_j)^{site(i)}$. By defining $C_j^{min(j)} \prec C_j^{site(i)}$ in S then $C_j^{min(j)} \prec R_i(X_j)^{site(i)}$ and so $C_j \prec R_i(X_j)$. Thus H can be defined as a history over T.

Transformation (2) on Definition 5 ensures that a transaction is committed as soon as it has been committed at the first replica. Finally, no restriction about the beginning of a transaction is imposed in this definition. Hence, this definition is valid for the most general case of non-blocking protocols. Although Assumptions 1 and 2 are included in Definition 5, they do not guarantee that the obtained 1C-schedule is a SI-schedule. This is best illustrated in the following example, where it is also shown how the 1C-schedule may be built from each site SI-schedules.

Example 3. *In this example two sites (A, B) and the next set of transactions T_1, T_2, T_3, T_4 are considered: $T_1 = R_1(Y) W_1(X), T_2 = R_2(Z) W_2(X), T_3 = R_3(X) W_3(Z), T_4 = R_4(X) R_4(Z) W_4(Y)$. Figure 1 illustrates the mapping described in Definition 5 for building a 1C-schedule from the SI-schedules seen in the different nodes I_m. T_2 and T_3 are locally executed at site A ($RS_2 \neq \emptyset$ and $RS_3 \neq \emptyset$) whilst T_1 and T_4 are executed at site B respectively. The writesets are afterwards applied at the remote sites. Schedules obtained at both sites are SI-schedules, i.e. transactions read the latest version of the committed data at each site. The 1C-schedule is obtained from Definition 5. For* *example, the commit of T_1 occurs for the 1C-schedule in the minimum of the interval between C_1^A and C_1^B and so on for the remaining transactions. In the 1C-schedule of Figure 1, T_4 reads X_1 and Z_3 but the X_2 version exists between both (since X_2 was installed at site A). T_1 and T_2, satisfying that $WS_1 \cap WS_2 \neq \emptyset$, are executed at both sites in the same order. As T_1 and T_2 are not executed in the same order with regard to T_3, the obtained 1C-schedule is neither SI nor GSI.*

5   1–Copy–GSI Schedules

The 1C-schedule H_t obtained in Definition 5 will be a GSI-schedule if it verifies the conditions given in Definition 4. The question is what conditions local SI-schedules, H_t^k, have to verify in order to guarantee that H_t is a GSI-schedule. Taking into account the ordering of conflicting transactions in GSI-equivalence, we consider the kind of protocols that guarantee the same total order of the commit operations for the transactions with write/write conflicts at every site. However, the execution of write/write conflicting transactions in the same order at all sites does not offer SI nor GSI, as it has been shown in Example 3. Therefore, it is also necessary to consider the need of reading from a consistent snapshot from the notion of GSI-equivalence; i.e. all update transactions must be committed in the very same order at all sites. As a result, since all replicas generate SI-schedules and their local snapshots have received the same sequence of updates, transactions starting at any site are able to read a particular snapshot, that perhaps is not the latest one, but that is consistent with those of other replicas.

Assumption 3. (Total Order of Committing Transactions). *For each pair T_i, $T_j \in T$, a unique order relation $c_i^k < c_j^k$ holds for all SI-schedules H_t^k with $k \in I_m$.*

The SI-schedules H_t^k have the same total order of committed transactions. Without loss of generalization, we consider the following total order in the rest of this section: $c_1^k < c_2^k < ... < c_n^k$ for every $k \in I_m$.

The aim of the next theorem is to prove that the 1C-schedules generated by any deferred update protocol that verifies Assumption 3 are actually GSI-schedules; i.e., they comply with all conditions stated in Definition 4. Whilst proving that a transaction always reads from the same snapshot in a particular time interval is easy, it is not trivial to prove that for a given transaction T_i there has not been any other transaction T_j that has impacted T_i and that has been committed whilst T_i was being executed. However, due to the total commit order an induction proof is possible, showing that the obtained 1C-schedule verifies all conditions in order to be a GSI-schedule.

Theorem 1. *Under Assumption 3, the 1C-schedule H_t is a GSI-schedule.*

This theorem formally justifies such protocols correctness and establishes that their resulting isolation level is GSI; the proof of it is given in [6]. Additionally, it is worth noting that Assumption 3 is a sufficient condition, but not necessary, for obtaining GSI. Despite this, replication protocols that comply with such an assumption are easily implementable. In the next section, we analyze how to relax this assumption while obtaining GSI schedules with non-blocking protocols.

6   Relaxing Assumptions

Assumption 3 (Total order of committing transactions) is very strong. It forces to install the same snapshots in the same order at every replica. Thus, Theorem 1 guarantees that the 1C-schedule H_t is a GSI-schedule. On the contrary, the total order of conflicting transactions is not enough to guarantee SI nor GSI (see Example 3) and it requires a stronger condition: it is needed that the snapshot gotten by a transaction at its delegate replica matches the 1C-schedule, actually being the latter a GSI-schedule. However, this fact does not necessarily oblige each replica to install the same snapshots as in the 1C-schedule. That is, if $R_i(X_j)$ belongs to H_t then $X_j \in (Snapshot(DB, H_t, b_i) \cap Snapshot(DB, H_t^{site(i)}, b_i))$. From what it has been depicted before, it is clear that if you want to relax Assumption 3, you have to provide some property that sets a relation between the reads-from relationship of a transaction in the 1C-schedule and the reads-from relationship of the transaction local schedule at its delegate site. In the next, we provide more relaxing assumptions to obtain a 1C-schedule providing GSI.

Assumption 4. *For each pair $T_i, T_j \in T$ with $WS_i \cap WS_j \neq \emptyset$, a unique order relation $c_i^k < c_j^k$ holds for all SI-schedule H_t^k with $k \in I_m$; and, if there is some transaction $T_p \in T$ such that $c_i^k < c_p^k < c_j^k$ holds for some site $k \in I_m$ then it holds for every $k \in I_m$.*

This assumption states that between two conflicting transactions their commit ordering is the same at every site. Moreover, it also states that between both transactions, there are the same subset of committed transactions; no matter the order in which they occur.

Example 4. *Let us suppose that there are two replicas and the set of transactions $\{T_1, T_2, T_3, T_4, T_5, T_6, T_7\}$, with $WS_1 \cap WS_4 \neq \emptyset$, $WS_3 \cap WS_7 \neq \emptyset$ and the rest do not conflict among each other. At the first site you can find the following local SI-schedule: $c_1^1 < c_2^1 < c_3^1 < c_4^1 < c_5^1 < c_6^1 < c_7^1$ whilst at the second site the derived SI-schedule can be: $c_1^2 < c_2^2 < c_3^2 < c_4^2 < c_6^2 < c_5^2 < c_7^2$. In the latter, the commit ordering of transactions T_5 and T_6 is different from the scheduling of the former.*

As it may be inferred, Assumption 4 becomes Assumption 3 whenever the pattern of transactions do not allow to reorder the commit of transactions. In Example 4, it cannot happen without violating Assumption 4 the following: $c_4^2 < c_3^2$. On the other hand, taking Assumption 4 to the extreme, if all transactions do not conflict among them any committing order can be obtained at each site. To limit these situations from making their appearance, it is needed to enforce to each transaction to read from the same snapshot like for each pair of transactions $T_i, T_j \in T$ with $WS_j \cap RS_i \neq \emptyset$: they verify that if $c_j < b_i$ in H_t then $c_j^{site(i)} < c_i^{site(i)}$ in $H_t^{site(i)}$ that is stated in the next assumption.

Assumption 5 (Compatible Snapshot Read). *Let H_t be a 1C-schedule of the database DB, for each $T_i \in T$ there exists $s_i \leq b_i$ such that if $R_i(X_j) \in H_t$ then $X_j \in (Snapshot(DB, H_t, s_i) \cap Snapshot(DB, H_t^{site(i)}, b_i))$.*

This last assumption means that each transaction reads data items that belong to a valid global snapshot from the 1C-schedule although their delegate site do not install the same snapshot version. On the other hand Assumption 4, it seems clear that a 1C-schedule serializes the execution of conflicting transactions.

Property 3. *Under Assumption 4, the 1C-schedule H_t verifies that for each pair T_i, $T_j \in T$: $\neg(T_j$ impacts T_i at s_i).*

Proof. By Assumption 2, at any site $k \in I_m$, for each pair $T_j^k, T_i^k \in T^k$: $\neg(T_j^k$ impacts T_i^k at b_i^k). That is, either $WS_j^k \cap WS_i^k = \emptyset$ or $\neg(b_i^k < c_j^k < c_i^k)$. In the first case, by definition of T_j and T_i, $WS_j \cap WS_i = \emptyset$; hence, $\neg(T_j$ impacts T_i at b_i). In the latter case, let us assume that $WS_j^k \cap WS_i^k \neq \emptyset$. Again, by definition of T_j and T_i, $WS_j \cap WS_i \neq \emptyset$. Hence, either $\neg(T_j^k$ impacts T_i^k at b_i^k) or $\neg(T_i^k$ impacts T_j^k at b_j^k). Thus, $c_i^k < b_j^k$ or $c_j^k < b_i^k$ holds. By Assumption 4, $c_i^k < c_j^k$ for all sites $k \in I_m$. Thus, $c_i^k < b_j^k$ for all $k \in I_m$. In particular, $c_i^{site(j)} < b_j^{site(j)}$. By definition of H_t: $c_i < c_j$ and $c_i \leq c_i^{site(j)} < b_j$ holds in H_t. Suppose that T_j impacts T_i at b_i in H_t. That is, $WS_j \cap WS_i \neq \emptyset$ and $b_i < c_j < c_i$. A contradiction with $c_i < c_j$ is obtained. Therefore, $\neg(T_j$ impacts T_i at b_i). Analogously, if T_i impacts T_j at b_j in H_t. That is, $WS_j \cap WS_i \neq \emptyset$ and $b_j < c_i < c_j$. A contradiction with $c_i < b_j$ is obtained again, and therefore, $\neg(T_i$ impacts T_j at b_j). $\qquad\square$

In the next theorem is proved that 1C-schedules generated by deferred update protocols following Assumption 4 and Assumption 5 verify Definition 4; i.e. they generate GSI schedules.

Theorem 2. *Under Assumption 4 and Assumption 5, the 1C-schedule H_t is a GSI-schedule.*

Proof. Firstly, notice that Assumption 4 implies total order of conflicting transactions. Given this total order of conflicting transactions, the 1C-schedule H_t satisfies for each $T_i \in T$ that $\neg(T_j$ impacts T_i at b_i) for every $T_j \in T$. Additionally, by Assumption 5, for each $T_i \in T$, if $R_i(X_j) \in H_t$ then $X_j \in (Snapshot(DB, H_t, s_i) \cap Snapshot(DB, H_t^{site(i)}, b_i^{site(i)}))$ with $s_i \in \mathbb{R}^+$ and $s_i \leq b_i$ (recall that $b_i = b_i^{site(i)}$). This fact makes true Condition (1) in Definition 4. Therefore, if $s_i = b_i$ for every $T_i \in T$ then Condition (2) in Definition 4 trivially holds. We need to prove Condition (2) in general. Thus, consider $s_i < b_i$; there must be a transaction $T_m \in T$ such that $s_i < c_m < b_i$ and $WS_m \cap RS_i \neq \emptyset$. Let T_m be the first transaction in H_t verifying such condition. Therefore, by Assumption 1 and Assumption 2 ($H_t^{site(i)}$ is a SI-schedule), $b_i^{site(i)} < c_m^{site(i)}$ holds. As $c_m < b_i$ then $c_m < c_i$ also holds. Assume that $WS_m \cap WS_i \neq \emptyset$, if $c_i^{site(i)} < c_m^{site(i)}$ then by Assumption 4 and construction of H_t, $c_i < c_m$ leading to a contradiction. So, $b_i^{site(i)} < c_m^{site(i)} < c_i^{site(i)}$. This implies, by Assumption 2 that $WS_m \cap WS_i = \emptyset$ and T_m verifies Condition (2) in Definition 4.

Every transaction $T_p \in T$ such that $s_i < c_p < c_m < b_i$ verifies that $WS_p \cap RS_i = \emptyset$ since T_m is the first one such that $WS_m \cap RS_i \neq \emptyset$. So, if $WS_p \cap WS_i \neq \emptyset$ then you can find $s_i' \in \mathbb{R}^+$: $s_i < c_p < s_i' < c_m < b_i$. At s_i', Assumption 5 is verified again for T_i by Definition 3 of snapshot. Furthermore, if $c_p < b_i$ then $c_p^{site(i)} < b_i^{site(i)} < c_i^{site(i)}$ due to Assumption 4 and construction of H_t (recall that $c_j = min_{k \in I_m}\{c_j^k\}$ for all T_j), $c_p < c_m < c_i$ in H_t that is a contradiction with the initial supposition of $c_m < c_p < b_i$. Thus, $WS_p \cap WS_i = \emptyset$ and Condition (2) in Definition 4 is verified for every transaction. The 1C-schedule is a GSI-schedule under the given assumptions. $\qquad\square$

If we proceed one step further, we can relax a little bit more Assumption 4. Let us assume that update transactions can be grouped in disjoint sets (hereinafter called *subsets*) where for any pair of transactions belonging to the same subset satisfy that their writeset intersection is empty; whereas transactions belonging to different subsets do not necessarily have an empty intersection. It is only required that these subsets are applied at all replicas (*Atomicity*) and in the same order in order to ensure the consistency of the replicated database. Finally, transactions belonging to the same subset can be applied at any order. In the following, we formalize this new relaxation.

Assumption 6. *Let $\mathcal{P}(T)$ be a partition of T such that for every subset $S_z \in \mathcal{P}(T)$, $S_z = \{T_i \in T : \forall T_j \in S_z, T_i \neq T_j, WS_i \cap WS_j = \emptyset\}$, and for any other $S_{z'} \in \mathcal{P}(T)$, with $z' \neq z$, $S_z \cap S_{z'} = \emptyset$. For each pair $S_z, S_{z'} \in \mathcal{P}(T)$, with $S_z \neq S_{z'}$, a unique order relationship $c^k_{last(S_z)} < c^k_{first(S_{z'})}$ holds for all SI-schedule H^k_t with $k \in I_m$; where $c^k_{last(S_z)}$ and $c^k_{first(S_{z'})}$ stands for the last commit time of of transactions in S_z and the first commit time of transactions in $S_{z'}$ respectively.*

If a given replication protocol guarantees that the same set of transactions are grouped in the same subset and these subsets are applied in the very same order (albeit transactions inside each subset can be applied in an arbitrary order) at every replica, then we can set up a synchronization point whenever a group of writesets have been applied so that Assumption 5 can be satisfied. It is easy to infer that if all these subsets are formed by a single transaction then this will be the case we have considered so far under Assumption 3 and by Theorem 1. On the other hand, if we do consider that conflicting transactions are included in different subsets with cardinality greater or equal to 1, provided that more non-conflicting transactions are added, then we will be under the scenario depicted in Assumption 4 and, hence, by Theorem 2 is held (along with Assumption 5). Let us see this with an example.

Example 5. *If we consider the set of transactions introduced in Example 4 with two replicas: $\{T_1, T_2, T_3, T_4, T_5, T_6, T_7\}$, with $WS_1 \cap WS_4 \neq \emptyset$, $WS_3 \cap WS_7 \neq \emptyset$ and the rest do not conflict among each other. Let us group transactions in the following subsets: $S_a = \{T_1, T_2\}$, $S_b = \{T_3, T_4\}$, $S_c = \{T_5, T_6, T_7\}$. Assume that the schedule generated at the first replica is $(c^1_1 < c^1_2) < (c^1_3 < c^1_4) < (c^1_5 < c^1_6 < c^1_7)$ while the schedule derived at the other one is: $(c^2_2 < c^2_1) < (c^2_4 < c^2_3) < (c^2_7 < c^2_6 < c^2_5)$. In other words, it has been applied, in the very same order, subsets $S_a < S_b < S_c$ in both schedules; however, the commit ordering of transactions contained in the same subset do differ.*

Form the previous example, we can prove (as the next theorem shows) that 1C-schedules derived from ROWA protocols satisfying Assumptions 5 and 6 do generate GSI schedules; i.e. they verify Definition 4.

Theorem 3. *Under Assumption 6 and Assumption 5, the 1C-schedule H_t is a GSI-schedule.*

Proof. Note that Assumption 6 totally orders the "reads-from" relationships between transactions and this will work no matter if transactions are linked together in subsets or not. Regarding to conflicting transactions Assumption 6 ensures the first half of Assumption 4: "*If $T_i \in S_z$ and $T_j \in S_{z'}$ ($z \neq z'$) such that $WS_i \cap WS_j \neq \emptyset$ then $c^k_i \leq c^k_{last(S_z)} < c^k_{first(S_{z'})} \leq c^k_j$ for all $k \in I_m$*". Thus, the issue relies on the second part of Assumption 4:"*$T_p \in T$ then $c^k_i < c^k_p < c^k_j$ for all $k \in I_m$*". However this is not a big deal since: either T_p does not belong to S_z nor $S_{z'}$ and, hence, Assumption 6 does imply Assumption 4; or, T_p does not conflict with T_i and is included in S_z so that $c_p^k < c_j^k$ or, otherwise, T_p does not conflict with T_j and is included in $S_{z'}$ so $c_i^k < c_p^k$, respectively for all $k \in I_m$.   □

From all discussed throughout this section, one can infer that a replication protocol that respects Assumption 4 and Assumption 5 will provide GSI to its executed transactions without needing to block transactions. The simplest, and most straightforward, solution is to define a conflict class [13,14] and each site is responsible for one (or several) conflict class. Thus, transactions belonging to different conflict classes will commit in any order at remote replicas while conflicting transactions belonging to the same conflict class are managed by the underlying DBMS of its delegate replica. Of course, this solution has its own pros and cons, we assume that each transaction exclusively belongs to a conflict class, i.e. no compound conflict classes, and it will read data and write data belonging to that class. However, it is a high application dependent and the granularity of the conflict class is undefined: it can range from coarse (at table level) to fine (at row level) granularity. On the other hand, if we jointly consider Assumptions 5 and 6 we can derive a replication protocol like the one considered in [15] where the order of committed transactions is known in advance according to the replica identifier. Replicas send messages that contain the set of locally committed transactions one after the other, according to their replica identifiers. Upon the delivery of this message to a replica, it will compare the set of local transactions ready to commit with the ones contained in the message. Hence, it will abort all local transactions that conflict and it will apply (and commit) the ones contained in the message. In the case of replicas's turn, the local transactions that had survived will be committed and their associated writesets sent to the rest of replicas joined in a single message. Thus, in this protocol remote transactions are grouped in subsets sorted by the replica identifier. Transactions belonging to the same subset can be committed in any order (satisfying Assumption 6) whereas local transactions can be started after one of these subsets has been applied (satisfying Assumption 5).

Finally, note that there exists a trade off between relaxing the conditions needed by a replication protocol [5,15] for achieving GSI schedules in a replicated setting and the potential blocking from starting until Assumption 5 is satisfied. Again, we are losing one of the advantages of GSI itself in a replicated setting against conventional SI. Thus, this issue is highly application dependent and if the client application is properly matched up with the replication protocol (or is predominately read-only), then the potential blocking of transactions can be harmless at all [13,14] and achieve a good performance.

7   Conclusions

It has been formalized the sufficient conditions to achieve 1-copy-GSI for non-blocking replication protocols following the deferred update technique that exclusively broadcast the writeset of transactions with SI replicas. They consist in providing global atomicity and applying (and committing) transactions in the very same order at all replicas. This means that there are other mechanisms to provide GSI in a replicated setting, some come at the cost of blocking the start of transactions [5] (which goes against the non-blocking nature of SI [1]) or by way of relaxing the total order of committed transactions given here. In particular, that between two conflicting transactions the same set of non-conflicting transactions must be committed and transactions started while applying in different order these writesets have read data items that belong to global valid versions. To sum up, all the properties that have been formalized in our paper seem to be assumed in some previous works, but none of them carefully identified nor formalized such properties. As a result, this paper provides a robust basis for designing and developing correct future replication protocols based on SI replicas.

References

1. Berenson, H., Bernstein, P.A., Gray, J., Melton, J., O'Neil, E.J., O'Neil, P.E.: A critique of ANSI SQL isolation levels. In: SIGMOD, pp. 1–10. ACM Press, New York (1995)
2. Fekete, A., Liarokapis, D., O'Neil, E., O'Neil, P., Shasha, D.: Making snapshot isolation serializable. ACM TODS 30(2), 492–528 (2005)
3. Elnikety, S., Pedone, F., Zwaenopoel, W.: Database replication using generalized snapshot isolation. In: SRDS, pp. 73–84. IEEE-CS, Los Alamitos (2005)
4. Plattner, C., Alonso, G., Özsu, M.T.: Extending dbmss with satellite databases. VLDB J. 17(4), 657–682 (2008)
5. Lin, Y., Kemme, B., Patiño-Martínez, M., Jiménez-Peris, R.: Middleware based data replication providing snapshot isolation. In: SIGMOD, pp. 419–430. ACM, New York (2005)
6. González de Mendívil, J.R., Armendáriz-Iñigo, J.E., Muñoz-Escoí, F.D., Irún-Briz, L., Garitagoitia, J.R., Juárez-Rodríguez, J.R.: Non-blocking ROWA protocols implement GSI using SI replicas. Technical Report ITI-ITE-07/10, ITI (2007)
7. Pedone, F.: The database state machine and group communication issues (N. 2090). PhD thesis, EPFL (1999)
8. Chockler, G., Keidar, I., Vitenberg, R.: Group communication specifications: a comprehensive study. ACM Comput. Surv. 33(4), 427–469 (2001)
9. Wiesmann, M., Schiper, A.: Comparison of database replication techniques based on total order broadcast. IEEE TKDE 17(4), 551–566 (2005)
10. Bernstein, P.A., Hadzilacos, V., Goodman, N.: Concurrency Control and Recovery in Database Systems. Addison Wesley, Reading (1987)
11. Papadimitriou, C.: The Theory of Database Concurrency Control. Computer Science Press (1986)
12. Armendáriz-Iñigo, J.E., Juárez-Rodríguez, J.R., González de Mendívil, J.R., Decker, H., Muñoz-Escoí, F.D.: K-bound GSI: a flexible database replication protocol. In: SAC, pp. 556–560. ACM, New York (2007)
13. Patiño-Martínez, M., Jiménez-Peris, R., Kemme, B., Alonso, G.: Consistent database replication at the middleware level. ACM TOCS 23(4), 375–423 (2005)
14. Amza, C., Cox, A.L., Zwaenepoel, W.: Conflict-aware scheduling for dynamic content applications. In: USENIX (2003)
15. Juárez-Rodríguez, J.R., Armendáriz-Iñigo, J.E., González de Mendívil, J.R., Muñoz-Escoí, F.D.: A database replication protocol where multicast writesets are always committed. In: ARES, pp. 120–127. IEEE-CS, Los Alamitos (2008)

Measuring the Usability of Augmented Reality e-Learning Systems: A User–Centered Evaluation Approach

Costin Pribeanu, Alexandru Balog, and Dragoş Daniel Iordache

ICI Bucharest, Bd. Mareşal Averescu No. 8-10, Bucureşti, Romania
{pribeanu,alexb,iordache}@ici.ro

Abstract. The development of Augmented Reality (AR) systems is creating new challenges and opportunities for the designers of e-learning systems. The mix of real and virtual requires appropriate interaction techniques that have to be evaluated with users in order to avoid usability problems. Formative usability aims at finding usability problems as early as possible in the development life cycle and is suitable to support the development of such novel interactive systems. This work presents an approach to the user-centered usability evaluation of an e-learning scenario for Biology developed on an Augmented Reality educational platform. The evaluation has been carried on during and after a summer school held within the ARiSE research project. The basic idea was to perform usability evaluation twice. In this respect, we conducted user testing with a small number of students during the summer school in order to get a fast feedback from users having good knowledge in Biology. Then, we repeated the user testing in different conditions and with a relatively larger number of representative users. In this paper we describe both experiments and compare the usability evaluation results.

Keywords: User-centered design, Usability, Formative usability evaluation, Augmented reality, e-Learning.

1 Introduction

The development of Augmented Reality (AR) systems is challenging designers with new interaction paradigms seeking to take advantage by the broad range of possibilities in mixing real and digital environments. Real objects became part of the interaction space thus being used as versatile interaction objects which are playing various roles.

This paper aims at presenting an approach to the user-centered formative usability evaluation of an interaction scenario for AR-based educational systems developed in the framework of the ARiSE (Augmented Reality for School Environments) research project.

The main objective of the ARiSE project is to test the pedagogical effectiveness of introducing AR in schools and creating remote collaboration between classes around AR display systems. ARiSE will develop a new technology, the Augmented Reality Teaching Platform (ARTP) in three stages thus resulting three research

J. Cordeiro et al. (Eds.): ICSOFT 2008, CCIS 47, pp. 175–186, 2009.
© Springer-Verlag Berlin Heidelberg 2009

prototypes. Each prototype is featuring a new application scenario based on a different interaction paradigm. The first prototype implemented a Biology learning scenario for secondary schools. The implemented paradigm is 3D process visualization and is targeted at enhancing the students' understanding and motivation to learn the human digestive system.

In order to get a fast feedback from both teachers and students, each prototype is tested with users during the ARiSE Summer School which is held yearly. From each school, one teacher and 4 students (2 boys and 2 girls) are participating that are selected by the teacher based on their communication skills including English language speaking, and knowledge in the target discipline. Given this selection criteria, they are not so representative for the user population.

A first version of the Biology scenario has been developed in 2006 and tested with users during the 1st ARiSE Summer School which has been held in Hamrun, Malta. Since the usability evaluation results were not satisfactory, the interaction techniques have been re-designed and tested again during the 2nd ARiSE Summer School in October 2007 which has been held in Bucharest, Romania.

The basic idea of our approach was to conduct user testing during the summer school in order to get a fast feedback from users having good knowledge in Biology and to repeat the user testing in different conditions and with a relatively larger number of representative users. This means to perform formative evaluation in two stages and to analyze and compare results.

During both experiments, effectiveness and efficiency measures have been collected in a log file. Then a usability questionnaire has been administrated that is providing with both quantitative and qualitative measures of the educational and motivational values of the new learning scenario.

The rest of this paper is organized as follows. In the next section we present some related work in the area of usability evaluation of AR systems. The experiments are described in Section 3. Then we compare and discuss the similarities and differences between the evaluation results of both experiments. The paper ends with conclusion and future work in section 5.

2   Related Work

The design of user interfaces is part of the overall design of a software system. The evolution of HCI in the last decade shows a maturation of the discipline and a closer integration of usability evaluation methods in the iterative design process. This trend is mirrored someway by the evolution of two concepts which are relevant for user interface evaluation: usability and quality of software systems.

In the former version of ISO 9126 standard, usability has been defined as a software quality attribute that bears on the capability of being easy to understand, learn and operate with. Later on, the ISO standard 9241-11:1994 took a broader perspective on usability as the extent to which a product can be used by specified users to achieve specified goals effectively, efficiently and with satisfaction in a specified context of use. In this standard, the context of use has four main components: user, tasks, platform and environment [11].

These definitions were revised and integrated in the new version of ISO 9126 standard on quality of software systems [12] as follows:

- Usability is the capability of a software system to be understood, learned, used, and attractive to the user, when used under specified conditions.
- Quality in use is the extent to which a product used by specified users meets their needs to achieve specific goals with effectiveness, productivity, security and satisfaction in a specified context of use.

There are several approaches to usability evaluation and, consequently many usability evaluation methods [9]. In the last decade, many usability studies compared the effectiveness of various usability evaluation methods. As Hvannberg and Law pointed out [10], the trend is to delineate the trade-offs and to find ways to take advantage from the complementarities between different methods.

Depending on the purpose, the usability evaluation can be formative or summative. Formative usability testing is performed in an iterative development cycle and aims at finding and fixing usability problems as early as possible [18]. The earlier these problems are identified, the less expensive is the development effort to fix them. This kind of usability evaluation is called "formative" in order to distinguish from "summative" evaluation which is usually performed after a system or some component has been developed [15]. Summative usability evaluation is carried on by testing with a relatively large number of representative users and aims at finding strengths and weaknesses as well as comparing alternative design solutions or similar systems.

Formative usability evaluation can be carried on by conducting an expert-based usability evaluation (sometimes termed as heuristic evaluation) and / or by conducting user testing with a small number of users. In this last case, the evaluation is said to be user-centered, as opposite to expert-based formative evaluation. As Gabbard et al. [7] pointed out, this kind of user-based statistical evaluation can be especially effective to support the development of novel systems as they are targeted at a specific part of the user interface design.

Usability evaluation is important but it is not enough for users to accept a system. Usefulness and attitude towards the systems are also important. A well known model aiming to predict technology acceptance once users have the opportunity to test the system is TAM – Technology Acceptance Model [5]. TAM theory holds that use is influenced by user's attitude towards the technology, which in turn is influenced by the perceived ease of use and perceived usefulness. As Dillon & Morris pointed out [6], TAM provides with early and useful insights on whether users will or will not accept a new technology.

TAM is nowadays widely used as an information technology acceptance model. TAM has been tested to explain or predict behavioral intention on a variety of information technologies and systems, such as: word processors, spreadsheet software, email, graphics software, net conferencing software, online shopping, online learning, Internet banking and so on [16], [19].

Augmented reality systems are expensive and require a lot of research and design effort to develop visualization and rendering software. On another hand, the mix of real and virtual requires appropriate interaction techniques. According to Hix et al.

[8], the interaction components of AR systems are often poorly designed, thus reducing the usability of the overall system.

Designing for usability is not an easy task in the AR field, given the lack of both specific user-centered design methods and usability data [1], [2], [3], [4], [7]. The survey elaborated three years ago by Swan & Gabbard revealed that only 38 out of 1104 articles (published in four main AR publications during 1992-2004) addressed some aspect of HCI and only 21 (~2%) described a user testing experiment [17]. Despite the proliferation of AR-based systems, there is still little usability data available.

The situation is similar in the area of usability evaluation of AR-based e-learning systems. There are few studies reporting usability evaluation of e-learning systems targeting students from primary and secondary schools. Most of the existing usability studies are reporting either usability problems identified with expert-based methods or subjective measures of user satisfaction. Few usability studies are reporting measures of effectiveness and efficiency.

3   Evaluation Method and Procedure

3.1   Equipment

ARTP is a "seated" AR environment: users are looking to a see-through screen where virtual images are superimposed over the perceived image of a real object placed on the table [20]. In our case, the real object is a flat torso of the human body showing the digestive system. The test was conducted on the platform of ICI Bucharest. The real object and the pointing device could be observed in Figure 1. As it could be observed, two students staying face-to-face are sharing the same torso.

Fig. 1. Students testing the Biology scenario during the Summer School

A pointing device having a colored ball on the end of a stick and a remote controller Wii Nintendo as handler has been used as interaction tool that serves for three types of interaction: pointing on a real object, selection of a virtual object and selection of a menu item. The tasks as well as user guidance during the interaction were presented via a vocal user interface in the national language of students.

3.2 Participants and Tasks

3.2.1 Evaluation During Summer School

The 2nd ARiSE summer school was held in Bucharest on 24-28 October 2007. Two groups of 4 students and two teachers from German and Lithuanian partner schools together with three groups of 4 students accompanied by a total of 4 teachers from 3 general (basic) schools in Bucharest participated to the summer school. Testing and debriefing with users was done in the morning while the afternoon was dedicated for discussion between research partners.

20 students from which 10 boys and 10 girls tested the platform. None of the students was familiar with the AR technology. 12 students were from 8th class (13-14 years old), 4 from 9th class (14-15 years old) and 4 from 10th class (15-16 years old). Students have different ages because of the differences related to the curricula in each country.

The participants were assigned 4 tasks: a demo program explaining the absorption / decomposition process of food and three exercises: the 1st exercise asking to indicate the organs of the digestive system and exercises 2 and 3, asking to indicate the nutrients absorbed/ decomposed in each organ respectively the organs where a nutrient is absorbed/ decomposed.

3.2.2 Evaluation after the Summer School

Two classes (8th class), each from a different school in Bucharest participated at user testing in the period 1-15 November 2007. The total number of participants was 42 students from which 19 boys and 23 girls. None of the students was familiar with the AR technology. Students came in groups of 6-8 accompanied by a teacher, so testing has been organized in 2 sessions. The test has been conducted on the platform of ICI Bucharest.

The students have been assigned 3 tasks: a demo lesson, the 1st exercise and one of the exercises 2 or 3. The number of tasks assigned to a student has been reduced to 3, because of time limitations. After finishing the assigned exercises, students were free to perform the third exercise or to repeat an assigned one.

3.3 Method and Procedure

In order to meet the ARiSE project goals we took a broader view on usability evaluation by grounding our work in a TAM model. A usability questionnaire has been developed that is based on existing user satisfaction questionnaires, usability evaluation approaches and results from the 1st ARiSE Summer School in 2006.

The questionnaire has 28 closed items (quantitative measures) and 2 open questions, asking users to describe the most 3 positive and most 3 negative aspects (qualitative measures). The closed items are presented in Table 1.

This evaluation instrument provides with 24 items that are targeting various dimensions such as ergonomics, usability, perceived utility, attitude and intention to use. The remainder four items are to assess how the students overall perceived the platform as being easy to use, useful for learning, enjoyable to learn with and exciting. By addressing issues like perceived utility, attitude and intention to use, usability evaluation results could be easier integrated with pedagogical evaluation results.

Before testing, a brief introduction to the AR technology and ARiSE project has been done for all students. Then, each team tested the ARTP once, during 1 hour. Students were asked to watch the demo lesson and then to perform the three exercises in order.

Table 1. The usability evaluation questionnaire

No.	Item
1	Adjusting the "see-through" screen is easy
2	Adjusting the stereo glasses is easy
3	Adjusting the headphones is easy
4	The work place is comfortable
5	Observing through the screen is clear
6	Understanding how to operate with ARTP is easy
7	The superposition between projection and the real object is clear
8	Learning to operate with ARTP is easy
9	Remembering how to operate with ARTP is easy
10	Understanding the vocal explanations is easy
11	Reading the information on the screen is easy
12	Selecting a menu item is easy
13	Correcting the mistakes is easy
14	Collaborating with colleagues is easy
15	Using ARTP helps to understand the lesson more quickly
16	After using ARTP I will get better results at tests
17	After using ARTP I will know more on this topic
18	The system makes learning more interesting
19	Working in group with colleagues is stimulating
20	I like interacting with real objects
21	Performing the exercises is captivating
22	I would like to have this system in school
23	I intend to use this system for learning
24	I will recommend to other colleagues to use ARTP
25	Overall, I find the system easy to use
26	Overall, I find the system useful for learning
27	Overall, I enjoy learning with the system
28	Overall, I find the system exciting

During testing, effectiveness (binary task completion and number of errors) and efficiency (time on task) measures have been collected in a log file. Measures were collected for all exercises performed.

After testing, the students were asked to answer the new usability questionnaire by rating the items on a 5-point Likert scale (1-strongly disagree, 2-disagree, 3-neutral, 4-agree, and 5-strongly agree). Prior to the summer school, the questionnaire has been translated in the native language of students.

4   Evaluation Results and Comparison

4.1   Answers to the Questionnaire

Reliability of the scale was 0.93 (Cronbach's Alpha) for the first sample, respectively 0.95 for the second, which is acceptable. Overall, the results were acceptable since means are over 3.00 (i.e. neutral). A comparison with the summer school evaluation results is presented in Fig. 2.

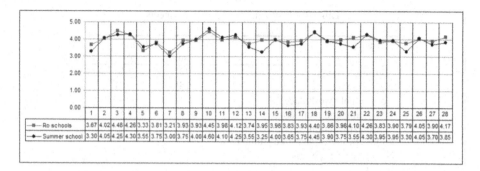

Fig. 2. Comparison with the summer school evaluation results

The general pattern is similar, in that items which were scored low at the summer school were also scored low by the students from the Romanian schools. In general, students participating to the summer school scored lower than students from the Romanian schools (general mean of 3.85 vs. 3.96).

For the three items below, the differences are relatively high (over 0.40) and showing deviations from the general pattern:

- Item 14 (Collaborating with colleagues is easy)
- Item 21 (Performing the exercises is captivating)
- Item 25 (Overall I find the system easy to use)

An independent samples t-test revealed that differences are statistically significant ($\alpha=0.05$, DF=60) only for the items: 14 (t=-2.164, p=0.034), and 21 (t=-2.231, p=0.029).

Analysis of the questionnaire data shows that two items were scored bellow 3.50 in both samples. These items (5 and 6) are targeting the clarity of visual perception. Also, in both samples four items were rated over 4.25: item 4 – the workplace is comfortable, item 10 – usefulness of the multimodal interaction in AR environments, item 18 – motivational value of the ARTP, and item 22 – intention to use, denoting an overall acceptance of the AR technology.

4.2 Most Mentioned Positive and Negative Aspects

The answers to the open questions have been analyzed in order to extract key words (attributes). Attributes have been then grouped into categories. Some students only described 1 or two aspects while others mentioned several aspects in one sentence thus yielding a number of 82 positive aspects and 69 negative aspects from the first sample respectively 85 positive aspects and 79 negative aspects from the second sample.

Main categories of most mentioned positive aspects are summarized in Table 2 in a decreasing order of their frequency.

Table 2. Most mentioned positive aspects

Category	After / during summer school	
Educational support	33	40
AR and 3D visualization	15	13
Comfortable workplace	11	-
Interesting and motivating	8	8
Vocal explanation	8	7
Funny, provocative (alike games)	7	7
Novel, good experience	-	4
Easy to use	3	3
Total	85	82

Educational support includes aspects like: better understanding ("*you understand better the real position of the organs*", "*the system help to better understand the lesson*", "*it helps you to understand where and how are the organs placed*"), good for learning ("*I learn easily the place of each organ*", "*the system helps me to learn better*"), easy to remember the lesson ("*I can better remember the learning content*"), attractive and faster learning ("*it is good for faster learning*"). Students from the Romanian schools also liked the exercises ("*very good exercises*").

These aspects correspond to the positive evaluation of items 4 (Using the application helps to understand the lesson more quickly) and 26 (Overall, I find the system useful for learning) in the usability questionnaire.

Students also liked the AR technology and 3D interaction ("*you learn the topic in 3D*"). Students liked the vocal explanation ("*explanations are good and descriptive*", "*I understood well the explanations*"). This is consistent with the positive evaluation of item 10 (Understanding the vocal explanation is easy) in the usability questionnaire.

Students also appreciated the AR system as funny ("*it was beautiful, like a game*"), novel and motivating ("*the system motivates to learn such topic*", "*the system makes learning more interesting*", "*it was interesting and captivating*"). These aspects are consistent with the positive evaluation of item 18 (The system makes learning more interesting).

Overall, the comparison with the summer school results is showing many similarities and small differences.

Most mentioned negative aspects are summarized in Table 3 in a decreasing order of their frequency.

Most frequent was the difficulty to reach each organ with the interaction tool. ("*it was often difficult to point to the right organ*", "*even if you know the right answer, is difficult to select it*", "*the pointer didn't select organs and sometimes didn't work*"). Selection and superposition problems as well as difficulties to use the system correspond to the low rating of items 7 (The superposition between projection and the real object is clear) and 25 (Overall, I find the system easy to use) in the usability questionnaire.

Second category of negative aspects was the eye pain provoked by the wireless glasses ("*it was something wrong with glasses. They were blinking*", "*the glasses were blinking*", "*after exercises we feel a pain in the eyes*").

Table 3. Most mentioned negative aspects

Category	After / during	
		summer school
Selection problems	25	25
Eye pain and problems with glasses	18	13
Real object too big	15	10
Superposition	7	4
Difficult to use	4	3
Other problems	10	14
Total	79	69

Many students complained about the fact that the real object was too big and it was difficult to work in pairs (*"I didn't like the fact that torso has to be moved"*, *"every student should have his own torso"*, *"I didn't like to move the torso with my colleague"*). The difference of frequency (15 vs. 10) corresponds to the different rating of item 14 (3.25 vs. 3.95) in the usability questionnaire.

Again, the comparison with the summer school evaluation results shows similar usability problems.

4.3 Measures of Effectiveness and Efficiency

A comparison between effectiveness and efficiency measures is presented in Table 4. Differences exist between the completion rates at the first and third exercise. Participants at the summer school made fewer errors. However, in both cases the third exercise was finished with many errors.

Differences exist for the number of errors and time on task between the two samples. An explanation is the fact that during summer school the participants had nothing else to do and the event itself was providing with an extra motivation (and some sense of competition) while students from Romanian schools came to user testing in the afternoon, after classes (they are learning in the morning) so they were already tired.

For the summer school participants, the first exercise was easier to solve (just to show organs) but more difficult to use. Errors (min=0, max=13, SD=3.9) are mainly due to the difficulties experienced with the selection. The execution time varied between 116 sec (2 errors) and 852 sec. (10 errors) with a mean of 381.8 sec (SD=218.1). The last two exercises were more difficult to solve (there is a many-to-many relationship between organs and nutrients). The second exercise was easier to use since the nutrients are selected with the remote controller. So we could infer that errors are mainly due to the lack of knowledge which is an argument for the pedagogical usefulness of the scenario. The execution time varied between 83 sec. (1 error) and 673 sec. (19 errors) with a mean of 254.9 sec. (SD=186.1).

4 students participating at the summer school failed to solve the third exercise. All students made errors. The execution time varied between 95 sec. (with 1 error) and 727 sec. (with 39 errors) with a mean execution time of 381.6 sec (SD=178).

Overall, 14 students succeeded to perform all the exercises at the summer school. The total execution time varied between 309 sec. (7 errors) and 1964 sec. (28 errors). The total number of errors varied between 6 and 56 errors with a mean of 23.3 errors. The total mean execution time was 1060 sec. i.e. 17.67 min. and is computed for the 14 students which succeeded to finish all the tasks.

Table 4. Measures of effectiveness and efficiency

Category	During summer school			After summer school		
	Rate	Errors	Time	Rate	Errors	Time
Exercise 1	100%	4.45	381.8	80%	6.88	455.8
Exercise 2	90%	4.94	254.9	91%	6.28	318.4
Exercise 3	80%	13.69	381.6	94%	15.90	401.4

For the students in the second sample, the first exercise has the lowest completion rate. 3 students from 35 failed to solve the second exercise. The execution time varied between 121 sec. (5 errors) and 932 sec. (6 errors) with a mean of 318.4 sec. (SD=220.1). One student from 17 failed to solve the third exercise. All students made errors and 7 students made over 20 errors. In this case, errors are due to the lack of knowledge and to the difficulties in selecting organs. The execution time varied between 174 sec. (with 3 errors) and 917 sec. (with 21 errors) with a mean execution time of 401.8 sec (SD=226.8).

Overall, 32 students (78%) in the second sample succeeded to perform all assigned exercises from which 11 students additionally performed a third exercise (not assigned). 6 students performed only one exercise while 3 students failed to perform any exercise. The total execution time for the 11 students performing all assigned exercises varied between 705 sec. (with 22 errors) and 1972 sec. (with 10 errors). The total number of errors varied between 8 and 50 with a mean of 20.73 errors (SD=12.73). The total mean time on task was 1207.8 sec. i.e. 20.1 min (SD=8.75).

5   Conclusions and Future Work

The evaluation of subjective measures of user satisfactions based on both quantitative and qualitative data collected with the usability questionnaire reveals several positive aspects.

ARTP has educational value: the system is good for understanding, good for learning, good for testing, and makes it easier to remember the lesson. The system makes learning faster. ARTP is increasing the students' motivation to learn: the system is attractive, stimulating and exciting, exercises are captivating and the system makes learning less boring. The students liked the interaction with 3D objects using AR techniques as well as the vocal explanation guiding them throughout the learning process.

Overall, the user acceptance of ARTP was good: students appreciated ARTP as useful for learning and expressed the interest to use it in the future.

Several usability problems exist that have been identified by both questionnaire data and log file analysis. The clarity of the visual perception should be improved as well as the overall ease of use. Many students complained about eye pains provoked by the wireless stereo glasses. Therefore it is strongly recommended to replace them with wired stereo glasses and to include this requirement into the technical specification of an AR educational platform.

Formative evaluation proved to be a useful aid to designers and a new version of the scenario has been recently released. By taking repeated measures on the same system version but with different user populations is both reliable for evaluators and convincing for designers.

The usability questionnaire is intended to support both formative and summative usability evaluation. In this respect, user testing performed after the summer school is also a first step to a summative evaluation of the Biology scenario. In order to gather enough data we restarted user testing in 2008, on an improved version of the ARTP.

Acknowledgements. We gratefully acknowledge the support of the ARiSE research project, funded under FP6-027039.

References

1. Bach, C., Scapin, D.: Obstacles and perspectives for evaluating mixed reality systems usability. In: Proceedings of Mixer Workshop - IUI-CADUI 2004, Funchal, pp. 72–79 (2004)
2. Bowman, D., Gabbard, J., Hix, D.: A Survey of Usability Evaluation in Virtual Environments: Classification and Comparison of Methods. Presence: Teleoperators and Virtual Environments 11(4), 404–424 (2002)
3. Costabile, M.F., De Angeli, A., Lanzilotti, R., Ardito, C., Buono, P., Pederson, T.: Explore! Possibilities and Challenges of Mobile Learning. In: Proc. of CHI 2008, pp. 145–154. ACM Press, New York (2008)
4. Coutrix, C., Nigay, L.: Mixed Reality: A Model of Mixed Interaction. In: Proceedings of Advanced Visual Interfaces, Venezia, pp. 59–64. ACM Press, New York (2006)
5. Davis, F.D., Bagozzi, R.P., Warshaw, P.R.: User Acceptance of Computer Technology: A Comparison of Two Theoretical Models. Management Science 35(8), 982–1003 (1989)
6. Dillon, A., Morris, M.: User acceptance of new information technology: theories and models. In: Williams, M. (ed.) Annual Review of Information Science and Technology, vol. 31, pp. 3–32 (1996); Medford NJ: Information Today
7. Gabbard, J., Hix, D., Edward Swan II, J.: User-Centered Design and Evaluation of Virtual Environments. IEEE Computer Graphics and Applications 19(6), 51–59 (1999)
8. Hix, D., Gabbard, J., Swan, E., Livingston, M., Herer, T., Julier, S., Baillot, Y., Brown, D.: A Cost-Effective Usability Evaluation Progression for Novel Interactive Systems. In: Proc. of Hawaii Intl. Conference on Systems Sciences, Track 9, p. 90276c. IEEE, Los Alamitos (2004)
9. Hornbaek, K.: Current practice in measuring usability: Challenges to usability studies and research. Intl. J. Human Computer Studies 64, 79–102 (2006)
10. Hvannberg, E.T., Law, L.-C.: Classification of usability problems (CUP) scheme. In: Proceedings of Interact 2003, Zurich, Switzerland, September 1-5 (2003)
11. ISO/DIS 9241-11:1994 Information Technology – Ergonomic Requirements for Office Work with visual display terminal (VDTs) - Guidance on usability (1994)
12. ISO 9126-1:2001 Software Engineering - Software product quality. Part 1: Quality Model (2001)
13. Kaufmann, H., Dunser, A.: Summary of Usability Evaluation of an Educational Augmented Reality Application. In: Shumaker, R. (ed.) HCII 2007 and ICVR 2007. LNCS, vol. 4563, pp. 660–669. Springer, Heidelberg (2007)
14. Jeon, S., Shim, H., Kim, G.: Viewpoint Usability for Desktop Augmented Reality. The International Journal of Virtual Reality 5(3), 33–39 (2006)
15. Scriven, M.: Evaluation thesaurus, 4th edn. Sage Publications, Newbury Park (1991)

16. Sun, H., Zhang, P.: The role of moderating factors in user technology acceptance. International Journal of Human-Computer Studies 64, 53–78 (2006)
17. Swann II, J.E., Gabbard, J.L.: Survey of User-Based Experimentation in Augmented Reality. In: Proceedings of 1st International Conference on Virtual Reality, Las Vegas, Nevada, July 22-27 (2005)
18. Theofanos, M., Quesenbery, W.: Towards the Design of Effective Formative Test Reports. Journal of Usability Studies 1(1), 27–45 (2005)
19. Venkatesh, V., Davis, F.D., Morris, M.G.: Dead or Alive? The Development, Trajectory And Future Of Technology Adoption Research. Journal of the AIS 8(4), 267–286 (2007)
20. Wind, J., Riege, K., Bogen, M.: Spinnstube®: A Seated Augmented Reality Display System. In: Virtual Environments, IPT-EGVE – EG/ACM Symposium Proceedings, pp. 17–23 (2007)

Supporting the Process Assessment through a Flexible Software Environment

Tomás Martínez-Ruiz[1], Francisco J. Pino[1,2], Eduardo León-Pavón[1], Félix García[1], and Mario Piattini[1]

[1] ALARCOS Research Group, Information Systems and Technologies Department
UCLM-INDRA Research and Development Institute, University of Castilla-La Mancha
Paseo de la Universidad, 4 – 13071, Ciudad Real, Spain
{Tomas.Martinez,Felix.Garcia,Mario.Piattini}@uclm.es
eduardo.leon@steria.es
[2] IDIS Research Group, Electronic and Telecommunications Engineering Faculty
University of Cauca, Street 5 # 4 – 70, Popayán, Colombia
fjpino@unicauca.edu.co

Abstract. A process assessment method provides elements in order to produce information which gives an overall view of the process capability. This information helps to determine the current state of software processes, their strengths and weaknesses. Owing to the fact that the assessment can be performed according to a given assessment process or any other and the processes of the organization can also use one particular process model or any other, we have developed an environment composed of two components; one of these generates the database schema for storing the process reference model and assessment information and the other one assesses the process with reference to this information, generating results in several formats, to make it possible to interpret data. The goal of this work is to show an environment that allows us to carry out assessments that are in accord with various different process assessment models, on several process reference models. This assessment environment has been developed and used in the context of the COMPETISOFT project in order to support its process assessment activities.

Keywords: Software process improvement, Software process assessment, Tools support, Assessment tool, COMPETISOFT.

1 Introduction

Quality is the most effective way to introduce any product into the buyers' market at the present time. Furthermore, due to the importance of software products in our daily life, quality is a decisive factor in guaranteeing that products are able to do their jobs properly. Software development organizations know the importance of this aspect and they are indeed interested in the quality of software products they create [1]. But the quality of a product depends on the capability of the processes in which this product is created [2]. Process capability is therefore an essential characteristic and there are two factors in this: one is image and the other is sheer need. They have to project a positive image if

J. Cordeiro et al. (Eds.): ICSOFT 2008, CCIS 47, pp. 187–199, 2009.

they are to export the software they produce and they need to turn their projects into effective and efficient ones.

Improvement of software processes is the way to maximize both factors. Three elements are needed when carrying out a process improvement initiative; a process improvement method, a process reference model and a process assessment method [3]. The process assessment method provides elements in order to produce qualitative and quantitative results which give an overall view of the process capability. These results give information that helps to determine the current state of software processes, their strengths and weaknesses. That information serves to define strategies for the execution of process improvement.

However, given the importance of assessing the processes before, during and after the improvement has been performed, several tools have been developed to help users assess processes. These tools can carry out repetitive actions, by reducing the cognitive charge of people involved in the assessments, and they can perform most of the management tasks that were done manually.

The diversity of existing software process models and assessment methods has led to the development of tools for the evaluation of the processes of each process model with reference to each assessment method. Each one of these tools depends on the appropriate process and assessment method.

In this paper we present an environment that is able to evaluate processes following any process reference model and any assessment method. How this environment supports the processes assessment of COMPETISOFT is also presented. This environment provides companies with the technical support necessary to carry out and store the results of their assessments in an integrated and consistent way. And it also avoids the development of specific tools for assess following each new assessment method or each new process reference model. With this environment, assessing software processes can be conducted automatically and in a flexible way by using a generic structure and assessment procedures. This avoids having to use diverse tools to conduct process assessments according to specific process reference models and assessment methods. Furthermore, EvalTOOL environment generates the information about assessments in a consistent and simple way, including graphics showing a summary with the assessment results and it also helps users to use them.

The rest of the paper is structured as follows. In section 2 the state of the art in software process and tools is analyzed. Section 3 presents the EvalTOOL environment and describes its main components. In section 4, the use of the environment is illustrated by means of some examples. In section 5 the support that the environment offers to the process assessment of COMPETISOFT is showed. Finally, the conclusions and future work are outlined.

2 State of the Art

Several process models and assessment models have been created in recent times. In this context it is important to emphasize next proposals:

- The Software Engineering Institute of Carnegie Mellon University has developed the *Capability Maturity Model Integrated* (CMMI) [4]. This model is based on CMM (Capability Maturity Model) and contains the best practices, grouped in several processes. This process model defines six maturity levels, which classify organizations in a range from chaotic level to continuous improvement level and it can be used in staged or continuous representations. CMMI uses the *Software CMMI Appraisal Method for Process Improvement* (SCAMPI) [5]. This method includes the best assessment practices and defines three steps to plan and prepare the assessment, carry it out, and inform of its results.

- The International Organization for Standardization has defined both a process and an assessment method: ISO 12207 [6] is a life cycle process model that defines the main activities that must be performed during the software development. It groups these activities into processes and categorizes the processes. This norm specifies the life cycle process architecture but not how to implement it. ISO 15504 [7-9] is the standard for carrying out assessments. The last version is divided into seven parts and it defines the minimum requirements to guarantee that the assessment results obtained are repeatable and consistent. It defines six capability levels.

- Regarding to models oriented to Small and Medium Enterprises (SMEs) we have developed the COMPETISOFT project [10]. The goal of this project is to increase the competitiveness of Ibero-American small software organizations, through the creation and dissemination of a common methodological framework for the processes improvement. The methodological framework of COMPETISOFT is composed of a process reference model, a process assessment method and an improvement framework. The process reference model defines ten processes, grouped into three categories as an organization hierarchy. With regard to the process assessment, COMPETISOFT defines a formal assessment method (six capability levels), moreover within the improvement framework a self-assessment methodology (three capability levels) is also defined. Both formal and self assessments are based on ISO 15504-2.

On the other hand, several tools have been developed to help in the assessing of processes. The most outstanding ones are based on CMM, CMMI and ISO 15504:

- CMM-Quest [11] allows us to evaluate the most important processes of an organization, to determine strengths and weaknesses. It assigns values to objectives, but it can not assess process practices.

- Appraisal Wizard [12] is based on SCAMPI, using CMMI as process model. It offers support for assigning values to the elements of the process (practices, objectives, etc). It allows the re-use of the results of an assessment in another later one. There is a light version of this tool called Appraisal Lite [12].

- Spice 1-2-1 [13] assigns values to base and generic practices. SPiCE Lite [14] offers a quick and efficient way to detect weaknesses and strengths of the process. It shows results as reports or via web and it has two running modes.

- Appraisal Assistant Beta [15] offers support for evaluating the maturity of an organization by the creation of user defined models, converting results from one framework to another, and generating reports about each assessment. Appraisal Assistant is now in a Beta version.

In literature we can find other assessment tools, even tools based on spreadsheets. Their functionality is similar to that of the above tools. The main drawback of these tools in comparison to the present work is that each one of these tools uses a specific process model and a specific assessment method (to see Table 1). In the context of the COMPETISOFT project and to support its process assessment the present work have been developed. Moreover, our objective is to provide companies with a flexible tool so they can assess their software processes by using different reference models and assessment methods but by using a single environment which facilitates comparison of results. Another important characteristic is that an important development effort has focused on the user interaction facility, by creating a usable GUI.

Table 1. Process models and assessment methods used by assessment tools

Tool	Reference Model	Assessment Method
CMM-Quest	CMMI-SE/SW continuous	ISO/IEC 15504
Appraisal Wizard	CMM, CMMI-SE/SW staged	SCAMPI
Appraisal Lite	and continuous	
SPiCE 1-2-1 SPiCE Lite	ISO/IEC 15504	ISO/IEC 15504:1998
Appraisal Assistant Beta	CMMI	ISO/IEC 15504, SCAMPI
EvalTOOL	Any	Any

3 EvalTOOL

A Flexible Environment for the capability assessment of Software Processes has been developed; it is called EvalTOOL. This environment has the following characteristics:

- It allows assessment using different assessment methods (ISO/IEC 15504, SCAMPI, COMPETISOFT's process assessment method).
- It is flexible: it allows the defining and addition of new processes by process reference models and new assessment methods, when the methods and processes are compatible with the environment core.
- It allows comparisons between the results of several assessments.
- It stores the models in their repository.
- Its reports show information in diagram form.
- It has a very usable interface, by means of enriched interfaces.

To support the assessments using several assessment methods and several process reference models, we have defined a generic metamodel (see Fig. 1). In this, the elements defined in outstanding process reference models such as ISO 12207, CMMI, COMPETISOFT and respective assessment methods (ISO 15504-2 and SCAMPI), have been taken into account. The environment uses this metamodel to guide the assessments. Because of this, the environment is flexible; that is, the environment is able to assess any processes (from a process reference model) that are defined in accord with this metamodel. That being so, the new process included in the environment must be based on conformance of the process reference model with the metamodel defined in the core of the environment. In addition, some modules can be designed to implement the other assessment methods.

The environment is prepared to generate and store the results of each assessment in the same way. So it is possible to retrieve these data and compare them with the

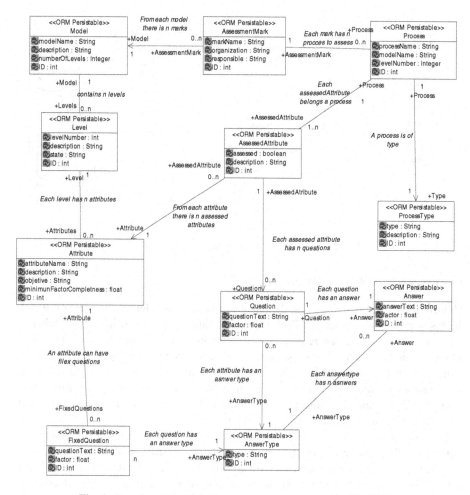

Fig. 1. Generic metamodel of process assessment used by EvalTOO

results of later assessments. In this way it is easy to see the improvement executed in each process of the organization quickly, by means of the comparison of the results of the process assessment. These results are shown using diagrams, to make them more understandable. Another feature is that its interface has been produced using enriched interfaces, so it is very easy and usable.

The environment is composed of two parts. The first one manages the process reference models and the second one applies the assessment methods over these processes. Fig. 2 shows the relationship between both parts which are described below. Both components are linked by the database, which is used by first one to write the new schemas. These schemas are used by the evaluation component to obtain information about the organization that is assessed and its processes.

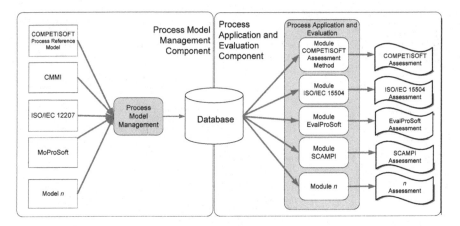

Fig. 2. Components of the environment EvalTOOL

3.1 Process Model Management

This component (left side of Fig. 2) manages the process reference models on which the assessments are conducted. This component supports the inclusion of flexibility in the environment processes related to different process reference models. This part allows the management of models, metamodels and schemas, by applying QVT transformations. It is able to generate the schema associated with each model, to allow the information of each process to be stored in a way that is compatible with the environment. The elements of the process (among others purpose, activities, roles and work products) stored are used to assess that process.

This part also defines the inverse transformation, from schema to the associated metamodel. The two transformations can be carried out using both run modes, automatically, where models or schemas are transformed without asking the user. They can also be customizable, in which the user contributes with the semantic information to achieve idempotent transformations between models and schemas.

This part is also able to generate a XMI file with the information stored in the base. To do this, the metamodel associated with the schema is obtained and used to create the XMI:SCHEMA.

The interface of the process model management is a desktop application and it is a very simple one. It offers help to the user and is available in English and Spanish. A more detailed description of this component can be found in [16].

3.2 Application and Evaluation Model

This component allows the definition and assessment of software process, using the databases created by the component described previously. It has been developed using the most advanced technologies. Because of this, it has a very simple, intuitive and easy GUI that helps users to manage the different tool functionalities (see Fig. 3).

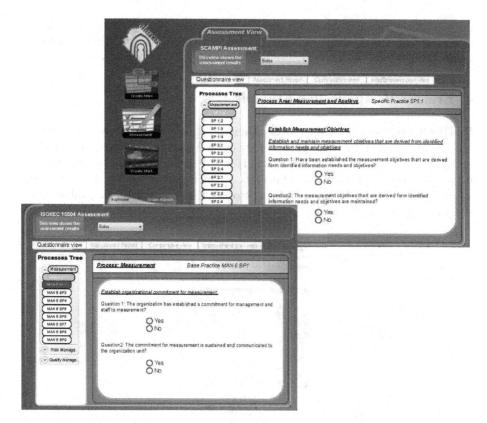

Fig. 3. Model application and evaluation part of EvalTOOL interface for SCAMPI and ISO/IEC 15504 assessments

It has the functionality of defining new assessment marks based on existing reference models, already included in the database. An assessment mark includes processes to be assessed and their evaluations. From this it is possible to add processes specific to an organization in order to evaluate them, with reference to one of the schemas defined as well as in accord with the metamodel used by the environment.

This evaluation part is further able to assess the chosen processes by means of answering several questions. The questions are defined from the process reference model stored in the database. An additional feature is that, due to the fact that the results of each assessment are stored in a common way, a comparison between them is possible. We can thereby obtain both the weaknesses and the strengths of processes (and organization). From these we know the points in which an improvement effort needs to be carried out. These are called *improvement opportunities*.

From the main menu (left side of Fig. 3) we can create and assess several marks. It is also possible to watch a demonstration of the application. This application of the environment has two versions; one of them is designed for PC, designed as a web page and offers all the functionality of these. The other one is designed to run in a Pocket PC and offers a subset of this functionality. This version is only able to answer the assessment questions about the process, in a dynamic way and within the scene where the software is developed and the processes take place. It allows us to do quick- assessment so the improvement opportunities can be known very rapidly.

4   EvalTOOL Application Example

EvalTOOL gives support to the process diagnosis activity of an improvement model. This environment has been designed to carry out software process assessments easily. In Fig. 4 we can see how the environment is related to the process reference models, the process assessment methods and the process improvement models. The aim of all this is to perform assessments of company process, whose results can be used to begin a process improvement cycle.

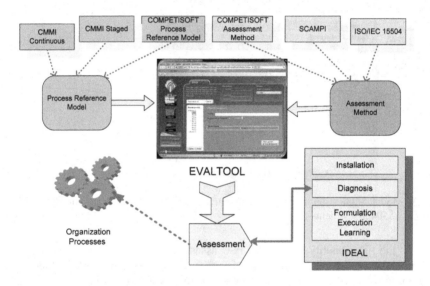

Fig. 4. Assessment program with EvalTOOL

Prior to assessment in the context of a software organization, some steps are necessary. First of all, the goals and benefits of the software process assessment must be presented to the organization work forces. Leaders must be in agreement with the assessment and the employees involved must receive the qualification about the models defined in the environment before they use it.

To do an assessment of a previously defined mark using EvalTOOL, it is only necessary to select an Evaluate Mark, and to choose the mark. Then the process can be assessed, by selecting and answering some questions about it. Questions are ordered by processes and their attributes. Once the questions are answered, you can see the assessment report generated (Fig. 5). The user is also informed about the process whose capability level is too low.

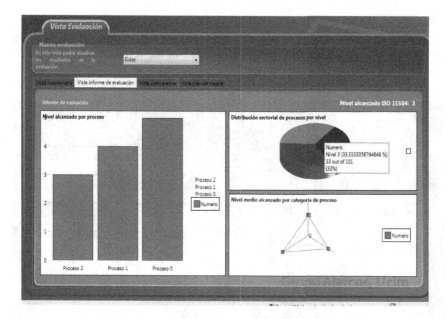

Fig. 5. EvalTOOL assessment results

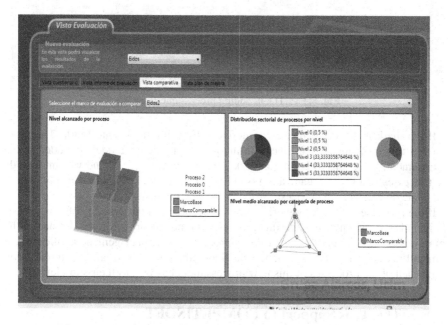

Fig. 6. EvalTOOL assessment result comparative

The environment stores the results obtained in each assessment and this information is very useful in seeing the progress between two assessments of the same improvement cycle. Fig. 6 shows a comparative chart of the results of two assessments carried out in the same improvement cycle.

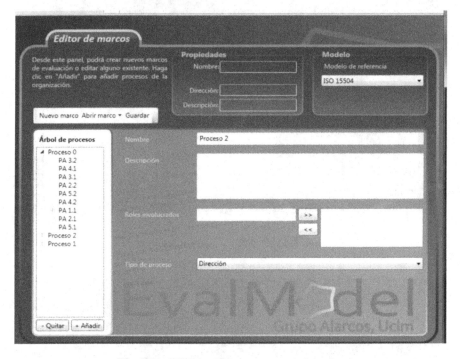

Fig. 7. EvalTOOL adding process questionnaire

If the mark is not defined, we can define it by using the Create Mark. The processes of this mark must be in line with the process reference models included in the environment. By means of adding a mark, a new mark is created, and we can include the processes of a given organization (see Fig. 7) within it. It is, moreover, possible to add some questions to the process attributes of each process.

If the processes of the new mark are not in accord with the process reference models included in the environment, this process reference model has to be defined in the environment. By means of the desktop process model management application, a compatible schema is created in the database, thereby storing all the information about each of the processes of this new process model as well as their sub-elements.

5   EvalTOOL as Support to COMPETISOFT

The methodological framework of COMPETISOFT is composed of a process reference model, a process assessment method and a framework for guiding the implementation of improvements (Improvement framework), see Fig. 8.

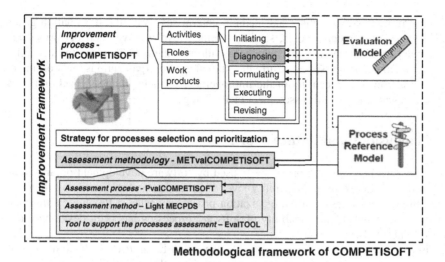

Methodological framework of COMPETISOFT

Fig. 8. Components of improvement framework of COMPETISOFT

Owing to the fact that the process assessment method, defined by this methodological framework, is conform to ISO 15504-2, the EvalTOOL environment supports directly this method.

On the other hand, the improvement framework of COMPETISOFT defines a process known as PmCOMPETISOFT [17] for guiding the ongoing process improvement. This process includes five activities: Initiating the cycle of improvement, diagnosing the process, formulating improvements, executing improvements and revising the cycle of improvement. In order to support the activity of diagnosing the software processes in small software organizations, we have developed a methodology for software process assessment called METvalCOMPETISOFT (see Fig. 8). This methodology allows us to obtain reliable information about the strengths and weaknesses of software processes, along with information on opportunities for improvement. The purpose is for that information to serve as a basis for decision making about process improvement within the organization. This methodology describes:

- A process for software processes assessment, called PvalCOMPETISOFT, which guides the activity of diagnosing the software processes step-by-step.
- An assessment method to determine the capability of software processes and maturity of a small organization, called Light MECPDS.
- A tool to support the process assessment, which is EvalTOOL.

The METvalCOMPETISOFT assessment methodology has been used in eight small organizations in order to carry out the processes assessment. The EvalTOOL environment has supported the application of the assessment methodology in order to carry out rapid assessments in these organizations, this type of assessment are a success factor for process improvements, as these were frequently applied with little time and few resources.

From the initial application in these organizations and bearing in mind the effort involved, it can be seen that the assessment methodology proposed is appropriate for diagnosing processes in small and medium enterprises. With the information generated by the assessment process, six improvement cycles have been performed and completed in six organizations, which has allowed them to increase the capability level of the processes assessed. Currently, the other two organizations are working on the formulating and executing of the improvement of their process, based on the information generated by this assessment methodology.

6 Conclusions and Future Work

The assessment of software processes is used by software development organizations as a means of knowing the capability level of processes and of improving them. To make the task of assessing the organization processes easier, a flexible environment for the assessment of the capability of software processes has been developed. It is called EvalTOOL.

The environment is designed in such a way as to be used to evaluate a software process by employing any assessment process. This flexibility allows organizations that have a lot of different projects and different processes to use a single tool to evaluate them. In addition, the environment offers the possibility of comparing two capability levels.

Furthermore, the evaluation of software processes gives the work forces of the organization the following positive features:

- Most of the questions presented in the questionnaires help users to have better knowledge of the changes that are being carried out, as well as the activities that would be performed and the difficulties entailed in them.
- The recommendations which the tool gives on the activities that are not carried out help them to provide feedback about their process improvement.
- The environment gives the improvement-manager a viewpoint from which to oversee the process assessed, over a period of time.

The environment is being used to carry out assessments based on the METvalCOM-PETISOFT assessment methodology. Several Iberoamerican organizations are using them to assess and improve their processes.

Our future work is to extend this environment by adding mechanisms to help users to improve their processes. Based on the results of the assessments, it is possible to determine the strengths and weaknesses of each process. That it turns makes it possible to establish what action is needed for the improvement of the process.

The environment is currently designed to run with a SQL Server database. Other future work can be to adapt it to run with other data sources, such as ODBC, or XML files.

We might also point out that the environment can be adapted to be used as a didactic tool. It can offer support for teaching users how to perform an assessment using any of the assessment methods and any process model.

Acknowledgements. This work is partially supported by the investigation into Software Process Lines sponsored by Sistemas Técnicos de Loterías del Estado S.A. within the framework of the agreement on the "Innovación del Entorno Metodológico de Desarrollo y Mantenimiento de Software", as well as by the projects: ESFINGE (TIN2006-15175-C05-05, Spanish Ministry of Science and Technology), COMPETI-SOFT (506AC0287, CYTED), MECENAS and INGENIO (PBI06-0024 and PAC08-0154-9262, Consejeria de Educación y Ciencia, JCCM, Spain).

References

1. Piattini, M., Garcia, F., Caballero, I.: Calidad de Sistemas Informáticos, Madrid: Ra-Ma, p. 388 (2006)
2. Fuggetta, A.: Software process: a roadmap. In: International Conference on Software Engineering (ICSE), pp. 25–34. ACM Press, New York (2000)
3. Pino, F., García, F., Piattini, M.: Software Process Improvement in Small and Medium Software Enterprises: A Systematic Review. Software Quality Journal 16(2), 237–261 (2008)
4. SEI, Capability Maturity Model Integration (CMMI), Version 1.1 CMMI (CMMI-SE/SW/IPPD/SS, V1.1) Staged Representation (2004)
5. SEI, Standard CMMI® Appraisal Method for Process Improvement (SCAMPI), Version 1.1: Method Definition Document (CMU/SEI-2001-HB-001) (2001)
6. ISO, ISO/IEC 12207:2002/FDAM 2. Information technology - Software life cycle processes (2004), http://www.iso.org
7. ISO, ISO/IEC 15504-2:2003/Cor.1:2004(E). Information technology - Process assessment - Part 2: Performing an assessment (2004), http://www.iso.org
8. ISO, ISO/IEC 15504-3:2003/Cor.1:2004(E). Information technology - Process assessment - Part 3: Guidance on Performing an Assessment (2004), http://www.iso.org
9. ISO, ISO/IEC 15504-4:2003/Cor.1:2004(E). Information technology - Process assessment - Part 4: Guidance on use for process improvement and process capability determination (2004), http://www.iso.org
10. Oktaba, H., Garcia, F., Piattini, M., Pino, F., Alquicira, C., Ruiz, F.: Software Process Improvement: The COMPETISOFT Project. IEEE Computer 40(10), 21–28 (2007)
11. CMM-Quest, http://www.cmm-quest.com/english (accessed in December 2008)
12. Appraisal Wizard,
    http://www.gantthead.com/sharedComponents/
    offsite.cfm?link=http%3A%2F%2Fwww.isd-inc.com
    (accessed in December 2008)
13. SPiCE 1-2-1, http://www.spice121.com/english (accessed in December 2008)
14. SPICE Lite, http://www.spicelite.com/English (accessed in December 2008)
15. Appraisal Assistant Beta,
    http://www.sqi.gu.edu.au/AppraisalAssistant/about.html
    (accessed in December 2008)
16. Martínez-Ruiz, T., García, F., Piattini, M.: Meta2Relational: Herramienta para la Gestión de Modelos de Procesos Software. In: VII JIISIC 2008. 2008. Guayaquil, Ecuador: Escuela Superior Politecnica del Litoral, pp. 251–258 (2008)
17. Pino, F., Vidal, J., Garcia, F., Piattini, M.: Modelo para la Implementación de Mejora de Procesos en Pequeñas Organizaciones Software. In: XII Jornadas de Ingeniería del Software y Bases de Datos, JISBD 2007, pp. 326–335 (2007)

Increasing Data Set Incompleteness May Improve Rule Set Quality

Jerzy W. Grzymala-Busse[1,2] and Witold J. Grzymala-Busse[3]

[1] Department of Electrical Engineering and Computer Science, University of Kansas
Lawrence, KS 66045, U.S.A.
[2] Institute of Computer Science, Polish Academy of Sciences, 01–237 Warsaw, Poland
[3] Touchnet Information Systems, Inc., Lenexa, KS 66219, U.S.A.

Abstract. This paper presents a new methodology to improve the quality of rule sets. We performed a series of data mining experiments on completely specified data sets. In these experiments we removed some specified attribute values, or, in different words, replaced such specified values by symbols of missing attribute values, and used these data for rule induction while original, complete data sets were used for testing. In our experiments we used the MLEM2 rule induction algorithm of the LERS data mining system, based on rough sets. Our approach to missing attribute values was based on rough set theory as well. Results of our experiments show that for some data sets and some interpretation of missing attribute values, the error rate was smaller than for the original, complete data sets. Thus, rule sets induced from some data sets may be improved by increasing incompleteness of data sets. It appears that by removing some attribute values, the rule induction system, forced to induce rules from remaining information, may induce better rule sets.

Keywords: Rough set theory, Rule induction, MLEM2 algorithm, Missing attribute values, Lost values, Attribute-concept values, "do not care" conditions.

1 Introduction

Recently data mining experiments with data sets affected by missing attribute values were reported in [1]. In these experiments we conducted a series of experiments on data sets that were originally complete, i.e., all attribute values were specified. First, for each data set, a portion of 10% of the total number of attribute values was replaced by special symbols denoting missing attribute values, or, in different words, this portion was replaced by missing attribute values. Then, with an increment of 10%, among remaining specified attribute values, a new portion of 10% was replaced by symbols of missing attribute values. This process was continued, with an increment of 10%, until all specified attribute values were replaced by symbols of missing attribute values. Then, for all data sets, error rates were computed using ten-fold cross validation. Obviously, during ten-fold cross validation experiments both training data and testing data, with the exception of original, complete data sets, were incomplete. It was observed that for some data sets an error rate was surprisingly stable, i.e., was not increasing as expected with increase of percentage of missing attribute values. Therefore we decided

J. Cordeiro et al. (Eds.): ICSOFT 2008, CCIS 47, pp. 200–216, 2009.

to perform different experiments of ten-fold cross validation in which training data sets are taken from incomplete data sets while testing data sets were taken from the original, complete data sets. In [1] we discussed three types of missing attribute values: *lost values* (the values that were recorded but currently are unavailable), *attribute-concept values* (these missing attribute values may be replaced by any attribute value limited to the same concept), and *"do not care" conditions* (the original values were irrelevant). A *concept* (class) is a set of all cases classified (or diagnosed) the same way.

Two special data sets with missing attribute values were extensively studied: in the first case, all missing attribute values are *lost*, in the second case, all missing attribute values are *"do not care" conditions*. Incomplete decision tables in which all attribute values are lost, from the viewpoint of rough set theory, were studied for the first time in [2], where two algorithms for rule induction, modified to handle lost attribute values, were presented. This approach was studied later, e.g., in [3,4], where the indiscernibility relation was generalized to describe such incomplete decision tables. In *attribute-concept values* interpretation of a missing attribute value, the missing attribute value may be replaced by any value of the attribute domain restricted to the concept to which the case with a missing attribute value belongs. For example, if for a patient the value of an attribute *Temperature* is missing, this patient is sick with *flu*, and all remaining patients sick with *flu* have values *high* or *very_high* for *Temperature* then using the interpretation of the missing attribute value as the *attribute-concept value*, we will replace the missing attribute value with *high* and *very_high*. This approach was studied in [5,6].

On the other hand, incomplete decision tables in which all missing attribute values are *"do not care" conditions*, from the view point of rough set theory, were studied for the first time in [7], where a method for rule induction was introduced in which each missing attribute value was replaced by all values from the domain of the attribute [7]. Such incomplete decision tables, with all missing attribute values being "do not care conditions", were broadly studied in [8,9], including extending the idea of the indiscernibility relation to describe such incomplete decision tables.

In this paper we report results of different experiments. In our new experiments, for every complete data set, we created a series of incomplete data sets by starting with a portion of 5% of the total number of attribute values. Then this portion of missing attribute values was incrementally enlarged, with an increment equal to 5% of the total number of missing attribute values.

Our new experiments started from creation, for every complete data set, a basic series of incrementally larger portions of missing attribute values with all missing attribute values being equal to "?" (lost values). For every basic series of data sets with missing attribute values, new series of data sets with missing attribute values were obtained by replacing all symbols of ? by symbols of "−" and"*", denoting different types of missing attribute values (attribute-concept values and "do not care" conditions). Additionally, the same basic series of data sets were used to induce certain and possible rule sets.

In general, incomplete decision tables are described by *characteristic relations*, in a similar way as complete decision tables are described by indiscernibility relations [10,11,12].

In rough set theory, one of the basic notions is the idea of lower and upper approximations. For complete decision tables, once the indiscernibility relation is fixed and the concept (a set of cases) is given, the lower and upper approximations are unique.

For incomplete decision tables, for a given characteristic relation and concept, there are three important and different possibilities to define lower and upper approximations, called singleton, subset, and concept approximations [10]. Singleton lower and upper approximations were studied, e.g., in [3,4,8,9]. Note that similar definitions of lower and upper approximations, though not for incomplete decision tables, were studied in [13,14,15]. Further definitions of approximations were discussed in [16,17]. Additionally, note that some other rough-set approaches to missing attribute values were presented in [5,7,18] as well.

Frequently, singleton approximations, both lower and upper, are not definable. So such approximations should not be used for data mining.

A preliminary version of this paper was presented at the Third International Conference on Software and data Technologies, ICSOFT'2008, Porto, Portugal, July 5–8, 2008 [19].

2   Blocks of Attribute-Value Pairs

We assume that the input data sets are presented in the form of a *decision table*. An example of a decision table is shown in Table 1.

Table 1. An example of the decision table

Case	Attributes Temperature	Headache	Cough	Decision Flu
1	very_high	no	no	no
2	?	no	yes	no
3	normal	yes	no	no
4	normal	yes	*	yes
5	–	yes	–	yes
6	very_high	?	yes	yes
7	normal	*	yes	yes
8	high	yes	yes	yes

Rows of the decision table represent *cases*, while columns are labeled by *variables*. The set of all cases will be denoted by U. In Table 1, $U = \{1, 2, ..., 8\}$. Independent variables are called *attributes* and a dependent variable is called a *decision* and is denoted by d. The set of all attributes will be denoted by A. In Table 1, $A = \{Temperature, Headache, Cough\}$. Any decision table defines a function ρ that maps the direct product of U and A into the set of all values. For example, in Table 1, $\rho(1, Temperature) = very\_high$. A decision table with an incompletely specified function ρ will be called *incomplete*.

For the rest of the paper we will assume that all decision values are specified, i.e., they are not missing. Also, we will assume that lost values will be denoted by "?" and "do not care" conditions by "*". Additionally, we will assume that for each case at least one attribute value is specified.

An important tool to analyze complete decision tables is a block of the attribute-value pair. Let a be an attribute, i.e., $a \in A$ and let v be a value of a for some case. For complete decision tables if $t = (a, v)$ is an attribute-value pair then a *block* of t, denoted $[t]$, is a set of all cases from U that for attribute a have value v. For incomplete decision tables the definition of a block of an attribute-value pair must be modified in the following way:

- If for an attribute a there exists a case x such that $\rho(x, a) = ?$, i.e., the corresponding value is lost, then the case x should not be included in any blocks $[(a, v)]$ for all values v of attribute a,
- If for an attribute a there exists a case x such that the corresponding value is a "do not care" condition, i.e., $\rho(x, a) = *$, then the case x should be included in blocks $[(a, v)]$ for all specified values v of attribute a.
- If for an attribute a there exists a case x such that the corresponding value is an attribute-concept value, i.e., $\rho(x, a) = -$, then the corresponding case x should be included in blocks $[(a, v)]$ for all specified values $v \in V(x, a)$ of attribute a, where

$$V(x, a) = \{\rho(y, a) \mid \rho(y, a) \text{ is specified}, y \in U, \rho(y, d) = \rho(x, d)\}.$$

For Table 1, $V(5, Temperature) = \{normal, high, very\_high\}$ and $V(5, Cough) = \{yes\}$, so the blocks of attribute-value pairs are:

[(Temperature, normal)] = $\{3, 4, 5, 7\}$,
[(Temperature, high)] = $\{5, 8\}$,
[(Temperature, very_high)] = $\{1, 5, 6\}$,
[(Headache, no)] = $\{1, 2, 7\}$,
[(Headache, yes)] = $\{3, 4, 5, 7, 8\}$,
[(Cough, no)] = $\{1, 3, 4\}$,
[(Cough, yes)] = $\{2, 4, 5, 6, 7, 8\}$,

For a case $x \in U$ the *characteristic set* $K_B(x)$ is defined as the intersection of the sets $K(x, a)$, for all $a \in B$, where the set $K(x, a)$ is defined in the following way:

- If $\rho(x, a)$ is specified, then $K(x, a)$ is the block $[(a, \rho(x, a)]$ of attribute a and its value $\rho(x, a)$,
- If $\rho(x, a) = ?$ or $\rho(x, a) = *$ then the set $K(x, a) = U$,
- If $\rho(x, a) = -$, then the corresponding set $K(x, a)$ is equal to the union of all blocks of attribute-value pairs (a, v), where $v \in V(x, a)$ if $V(x, a)$ is nonempty. If $V(x, a)$ is empty, $K(x, a) = U$.

For Table 1 and $B = A$,

$K_A(1) = \{1, 5, 6\} \cap \{1, 2, 7\} \cap \{1, 3, 4\} = \{1\}$,
$K_A(2) = \{1, 2, 7\} \cap \{2, 4, 6, 7, 8\} = \{2, 7\}$,
$K_A(3) = \{3, 4, 5, 7\} \cap \{3, 4, 5, 7, 8\} \cap \{1, 3, 4\} = \{3, 4\}$,

$$K_A(4) = \{3, 4, 5, 7\} \cap \{3, 4, 5, 7, 8\} = \{3, 4, 5, 7\},$$
$$K_A(5) = \{4, 5, 7, 8\},$$
$$K_A(6) = \{1, 5, 6\} \cap \{2, 4, 6, 7, 8\} = \{5, 6\},$$
$$K_A(7) = \{3, 4, 5, 7\} \cap \{2, 4, 6, 7, 8\} = \{4, 5, 7\},$$
$$K_A(8) = \{5, 8\} \cap \{3, 4, 5, 7, 8\} \cap \{2, 4, 6, 7, 8\} = \{5, 8\}.$$

Characteristic set $K_B(x)$ may be interpreted as the set of cases that are indistinguishable from x using all attributes from B and using a given interpretation of missing attribute values. Thus, $K_A(x)$ is the set of all cases that cannot be distinguished from x using all attributes. The characteristic relation $R(B)$ is a relation on U defined for $x, y \in U$ as follows

$$(x, y) \in R(B) \text{ if and only if } y \in K_B(x).$$

Thus, the relation $R(B)$ may be defined by $(x, y) \in R(B)$ if and only if y is indistinguishable from x by all attributes from B. The characteristic relation $R(B)$ is reflexive but—in general—does not need to be symmetric or transitive. Also, the characteristic relation $R(B)$ is known if we know characteristic sets $K_B(x)$ for all $x \in U$. In our example, $R(A) = \{(1, 1), (2, 2), (2, 7), (3, 3), (3, 4), (4, 3), (4, 4), (4, 5), (4, 7), (5, 4),$ $(5, 5), (5, 7), (5, 8), (6, 5), (6, 6), (7, 4), (7, 5), (7, 7), (8, 5), (8, 8)\}$. The most convenient way is to define the characteristic relation through the characteristic sets.

For decision tables, in which all missing attribute values are lost, a special characteristic relation was defined in [3], see also, e.g., [4]. For decision tables where all missing attribute values are "do not care" conditions a special characteristic relation was defined in [8], see also, e.g., [9]. For a completely specified decision table, the characteristic relation $R(B)$ is reduced to the indiscernibility relation.

3   Definability

For completely specified decision tables, any union of elementary sets of B is called a B-definable set [20]. Definability for completely specified decision tables should be modified to fit into incomplete decision tables. For incomplete decision tables, a union of some intersections of attribute-value pair blocks, where such attributes are members of B and are distinct, will be called B-*locally definable* sets. A union of characteristic sets $K_B(x)$, where $x \in X \subseteq U$ will be called a B-*globally definable* set. Any set X that is B-globally definable is B-locally definable, the converse is not true. For example, the set $\{7\}$ is A-locally definable since $\{7\} = [(Temperature, normal)] \cap [(Headache, no)]$. However, the set $\{7\}$ is not A-globally definable. On the other hand, the set $\{2\}$ is not even A-locally definable. Obviously, if a set is not B-locally definable then it cannot be expressed by rule sets using attributes from B. This is why it is important to distinguish between B-locally definable sets and those that are not B-locally definable.

4   Lower and Upper Approximations

For completely specified decision tables lower and upper approximations are defined on the basis of the indiscernibility relation [20,21]. Let X be any subset of the set U of all cases. The set X is called a *concept* and is usually defined as the set of all cases defined by a specific value of the decision. In general, X is not a B-definable set. However, set X may be approximated by two B-definable sets, the first one is called a B-*lower approximation* of X, denoted by $\underline{B}X$ and defined as follows

$$\{x \in U \mid [x]_B \subseteq X\},$$

where $[x]_B$ denotes an equivalence class of x with respect to the indiscernibility relation R associated with B. The second set is called a B-*upper approximation* of X, denoted by $\overline{B}X$ and defined as follows

$$\{x \in U \mid [x]_B \cap X \neq \emptyset\}.$$

The above shown way of computing lower and upper approximations, by constructing these approximations from singletons x, will be called the *first method*. The B-lower approximation of X is the greatest B-definable set, contained in X. The B-upper approximation of X is the smallest B-definable set containing X.

As it was observed in [20], for complete decision tables we may use a *second method* to define the B-lower approximation of X, by the following formula

$$\cup\{[x]_B \mid x \in U, [x]_B \subseteq X\},$$

and the B-upper approximation of x may be defined, using the second method, by

$$\cup\{[x]_B \mid x \in U, [x]_B \cap X \neq \emptyset\}.$$

Obviously, for complete decision tables both methods result in the same respective sets, i.e., corresponding lower approximations are identical, and so are upper approximations.

For incomplete decision tables lower and upper approximations may be defined in a few different ways. In this paper we suggest three different definitions of lower and upper approximations for incomplete decision tables, following [10,11,12]. Again, let X be a concept, let B be a subset of the set A of all attributes, and let $R(B)$ be the characteristic relation of the incomplete decision table with characteristic sets $K_B(x)$, where $x \in U$. Our first definition uses a similar idea as in the previous articles on incomplete decision tables [3,4,8,9], i.e., lower and upper approximations are sets of singletons from the universe U satisfying some properties. Thus, lower and upper approximations are defined by analogy with the above first method, by constructing both sets from singletons. We will call these approximations *singleton*. A singleton B-lower approximation of X is defined as follows:

$$\underline{B}X = \{x \in U \mid K_B(x) \subseteq X\}.$$

A singleton B-upper approximation of X is

$$\overline{B}X = \{x \in U \mid K_B(x) \cap X \neq \emptyset\}.$$

In our example of the decision table presented in Table 1 let us say that $B = A$. Then the singleton A-lower and A-upper approximations of the two concepts: $\{1, 2, 3\}$ and $\{4, 5, 6, 7, 8\}$ are:

$$\underline{A}\{1,2,3\} = \{1\},$$
$$\underline{A}\{4,5,6,7,8\} = \{5,6,7,8\},$$
$$\overline{A}\{1,2,3\} = \{1,2,3,4\},$$
$$\overline{A}\{4,5,6,7,8\} = \{2,3,4,5,6,7,8\}.$$

We may easily observe that the set $\{1, 2, 3, 4\} = \overline{A}\{1,2,3\}$ is not A-locally definable since in all blocks of attribute-value pairs cases 2 and 7 are inseparable. Thus, as it was observed in, e.g., [10,11,12], singleton approximations should not be used, theoretically, for data mining and, in particular, for rule induction. The second method of defining lower and upper approximations for complete decision tables uses another idea: lower and upper approximations are unions of elementary sets, subsets of U. Therefore we may define lower and upper approximations for incomplete decision tables by analogy with the second method, using characteristic sets instead of elementary sets. There are two ways to do this. Using the first way, a *subset* B-lower approximation of X is defined as follows:

$$\underline{B}X = \cup\{K_B(x) \mid x \in U, K_B(x) \subseteq X\}.$$

A *subset* B-upper approximation of X is

$$\overline{B}X = \cup\{K_B(x) \mid x \in U, K_B(x) \cap X \neq \emptyset\}.$$

Since any characteristic relation $R(B)$ is reflexive, for any concept X, singleton B-lower and B-upper approximations of X are subsets of the subset B-lower and B-upper approximations of X, respectively [12]. For the same decision table, presented in Table 1, the subset A-lower and A-upper approximations are

$$\underline{A}\{1,2,3\} = \{1\},$$
$$\underline{A}\{4,5,6,7,8\} = \{4,5,6,7,8\},$$
$$\overline{A}\{1,2,3\} = \{1,2,3,4,5,7\},$$
$$\overline{A}\{4,5,6,7,8\} = \{2,3,4,5,6,7,8\}.$$

The second possibility is to modify the subset definition of lower and upper approximation by replacing the universe U from the subset definition by a concept X. A *concept* B-lower approximation of the concept X is defined as follows:

$$\underline{B}X = \cup\{K_B(x) \mid x \in X, K_B(x) \subseteq X\}.$$

Obviously, the subset B-lower approximation of X is the same set as the concept B-lower approximation of X. A *concept* B-upper approximation of the concept X is defined as follows:

$$\overline{B}X = \cup\{K_B(x) \mid x \in X, K_B(x) \cap X \neq \emptyset\} =$$
$$= \cup\{K_B(x) \mid x \in X\}.$$

The concept upper approximations were defined in [13] and [14] as well. The concept B-upper approximation of X is a subset of the subset B-upper approximation of X

[12]. For the decision table presented in Table 1, the concept A-upper approximations are

$$\overline{A}\{1,2,3\} = \{1,2,3,4,7\},$$

$$\overline{A}\{4,5,6,7,8\} = \{3,4,5,6,7,8\}.$$

Note that for complete decision tables, all three definitions of lower approximations, singleton, subset and concept, coalesce to the same definition. Also, for complete decision tables, all three definitions of upper approximations coalesce to the same definition. This is not true for incomplete decision tables, as our example shows.

5   LERS and LEM2

The data system LERS (Learning from Examples based on Rough Sets) [22,23] induces rules from inconsistent data, i.e., data with conflicting cases. Two cases are conflicting when they are characterized by the same values of all attributes, but they belong to different concepts (classes). LERS handles inconsistencies using rough set theory, introduced by Z. Pawlak in 1982 [20,21]. In rough set theory lower and upper approximations are computed for concepts involved in conflicts with other concepts.

Rules induced from the lower approximation of the concept *certainly* describe the concept, hence such rules are called *certain* [24]. On the other hand, rules induced from the upper approximation of the concept describe the concept *possibly*, so these rules are called *possible* [24]. In general, LERS uses two different approaches to rule induction: one is used in machine learning, the other in knowledge acquisition. In machine learning, the usual task is to learn the smallest set of minimal rules, describing the concept. To accomplish this goal, LERS uses three algorithms: LEM1, LEM2, and MLEM2.

5.1   LEM2

The LEM2 option of LERS is most frequently used for rule induction since—in most cases—it gives better results than LEM1. LEM2 explores the search space of attribute-value pairs. Its input data set is a lower or upper approximation of a concept, so its input data set is always consistent. In general, LEM2 computes a local covering and then converts it into a rule set. We will quote a few definitions to describe the LEM2 algorithm [22,25,26].

The LEM2 algorithm is based on an idea of an attribute-value pair block. Let X be a nonempty lower or upper approximation of a concept represented by a decision-value pair (d, w). Set X *depends* on a set T of attribute-value pairs $t = (a, v)$ if and only if

$$\emptyset \neq [T] = \bigcap_{t \in T} [t] \subseteq X.$$

Set T is a *minimal complex* of X if and only if X depends on T and no proper subset T' of T exists such that X depends on T'. Let $\mathcal{T}$ be a nonempty collection of nonempty sets of attribute-value pairs. Then $\mathcal{T}$ is a *local covering* of X if and only if the following conditions are satisfied:

- each member T of $\mathcal{T}$ is a minimal complex of X,
- $\bigcup_{t \in \mathcal{T}}[T] = X$, and
- $\mathcal{T}$ is minimal, i.e., $\mathcal{T}$ has the smallest possible number of members.

MLEM2, a modified version of LEM2, processes numerical attributes differently than symbolic attributes. For numerical attributes MLEM2 sorts all values of a numerical attribute. Then it computes cutpoints as averages for any two consecutive values of the sorted list. For each cutpoint q MLEM2 creates two blocks, the first block contains all cases for which values of the numerical attribute are smaller than q, the second block contains remaining cases, i.e., all cases for which values of the numerical attribute are larger than q. The search space of MLEM2 is the set of all blocks computed this way, together with blocks defined by symbolic attributes. Starting from that point, rule induction in MLEM2 is conducted the same way as in LEM2.

Additionally, the newest version of MLEM2, used for our experiments, with merging intervals, at the very end simplifies rules by, as its name indicates, merging intervals for numerical attributes.

For Table 1, rules in the LERS format (every rule is equipped with three numbers, the total number of attribute-value pairs on the left-hand side of the rule, the total number of cases correctly classified by the rule during training, and the total number of training cases matching the left-hand side of the rule) [23] are:

certain rules, induced from the *concept* lower approximations:
2, 1, 1
(Temperature, very_high) & (Headache, no) -> (Flu, no),
2, 4, 4
(Cough, yes) & (Headache, yes) -> (Flu, yes),
2, 2, 2
(Temperature, very_high) & (Cough, yes) -> (Flu, yes),

and possible rules, induced from the *concept* upper approximations:
1, 2, 3
(Headache, no) -> (Flu, no),
1, 2, 3
(Cough, no) -> (Flu, no),
1, 2, 3
(Headache, yes) -> (Flu, yes),
2, 2, 2
(Temperature, very_high) & (Headache, yes) -> (Flu, yes).

5.2 LERS Classification System

Rule sets, induced from data sets, are used mostly to classify new, unseen cases. A classification system used in LERS is a modification of the well-known bucket brigade algorithm [27,28]. Some classification systems are based on a rule estimate of probability. Other classification systems use a decision list, in which rules are ordered, the first rule that matches the case classifies it [29]. In this section we will concentrate on a classification system used in LERS.

The decision to which concept a case belongs to is made on the basis of three factors: *strength*, *specificity*, and *support*. These factors are defined as follows: *strength* is the total number of cases correctly classified by the rule during training. *Specificity* is the total number of attribute-value pairs on the left-hand side of the rule. The matching rules with a larger number of attribute-value pairs are considered more specific. The third factor, *support*, is defined as the sum of products of strength and specificity for all matching rules indicating the same concept. The concept C for which the support, i.e., the following expression

$$\sum_{matching\ rules\ r\ describing\ C} Strength(r) * Specificity(r)$$

is the largest is the winner and the case is classified as being a member of C.

In the classification system of LERS, if complete matching is impossible, all partially matching rules are identified. These are rules with at least one attribute-value pair matching the corresponding attribute-value pair of a case. For any partially matching rule r, the additional factor, called *Matching_factor* (r), is computed. Matching_factor (r) is defined as the ratio of the number of matched attribute-value pairs of r with a case to the total number of attribute-value pairs of r. In partial matching, the concept C for which the following expression is the largest

$$\sum_{\substack{partially\ matching \\ rules\ r\ describing\ C}} Matching\_factor(r)* \\ Strength(r) * Specificity(r)$$

Note that our basic assumption was that for every case at least one attribute value should be specified. Thus, the process of enlarging the portion of missing attribute values was terminated when, during three different attempts to replace specified attribute values by missing ones, a case with all missing attribute values was generated.

6 Experiments

In our experiments we used seven typical data sets, see Table 2.

Table 2. Data sets used for experiments

| Data set | Number of | | |
	cases	attributes	concepts
Bankruptcy	66	5	2
Breast cancer	277	9	2
Hepatitis	155	19	2
Image segmentation	210	19	7
Iris	150	4	3
Lymphography	148	18	4
Wine	178	13	3

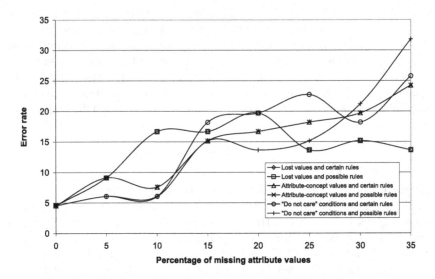

Fig. 1. Bankruptcy data set

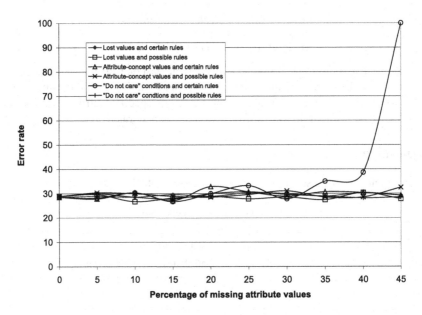

Fig. 2. Breast cancer data set

During experiments of ten-fold cross validation, training data sets were affected by incrementally larger and larger portion of missing attribute values, while testing data sets were always the original, completely specified data sets.

For any ten experiments of ten-fold cross validation all ten parts for both data sets: complete and incomplete were pairwise equal (if not taking missing attribute values

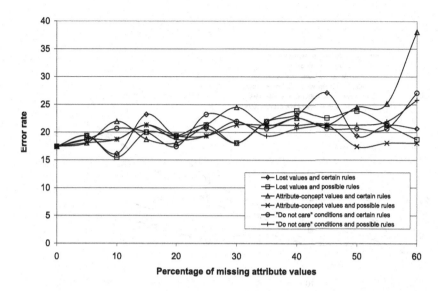

Fig. 3. Hepatitis data set

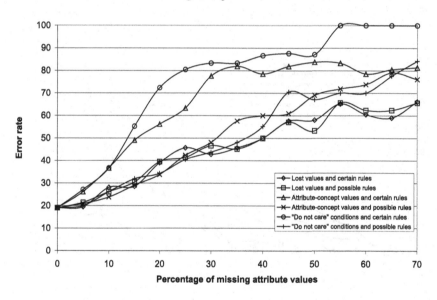

Fig. 4. Image segmentation data set

into account), i.e., any such two parts, complete and incomplete, would be equal if we will put back the appropriate specified attribute values into the incomplete part.

Results of experiments are presented in Figures 1–7. For some data sets (*bankruptcy* and *image*) the error rate increases rather consistently with our expectations: the larger percentage of missing attribute values the greater error rate. On the other hand, it is quite clear that for some data sets (*breast cancer*, *hepatitis*, *iris*, *lymphography* and *wine*) the

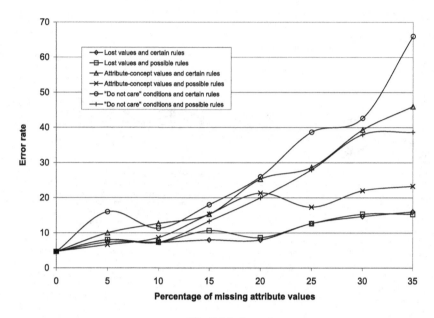

Fig. 5. Iris data set

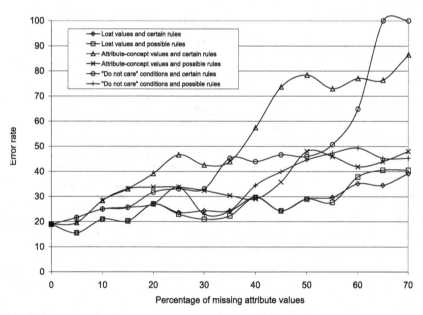

Fig. 6. Lymphography data set

error rate is approximately stable with increase of the percentage of missing attribute values of the type lost, while for some data sets (*breast cancer* and *hepatitis*) the error rate is stable for all three types of missing attribute values, except the largest percentage

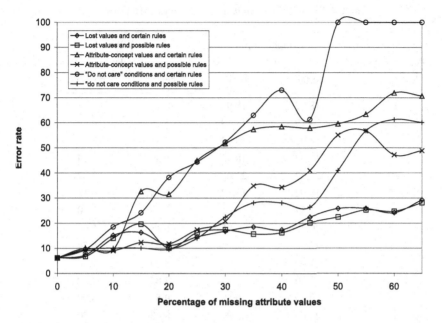

Fig. 7. Wine data set

Table 3. Error rate for *breast cancer*, *hepatitis*, and *iris* data sets

Percentage of lost values	Breast cancer rules		Hepatitis rules		Iris rules	
	certain	possible	certain	possible	certain	possible
0	28.52	28.88	17.42	17.42	4.67	4.67
5	29.6	29.6	18.06	18.06 5	4.0	4.67
10	28.88	28.52	15.48	15.48	6.0	5.33
15	29.6	29.24	20.65	16.77	6.0	6.0
20	27.8	29.24	18.06	18.06	6.0	4.67
25	28.52	29.6	19.35	19.3	6.67	7.33
30	28.88	27.8	19.35	19.35	6.67	7.33
35	27.44	27.8	17.42	17.42	6.67	6.67
40	29.24	28.52	18.71	18.71		
45	29.24	28.52	18.71	18.71		
50			18.71	20.0		
55			17.42	17.42		
60			18.06	18.71		

Table 4. Error rate for *lymphography* and *wine* data sets

Percentage of lost values	Lymphography rules		Wine rules	
	certain	possible	certain	possible
0	18.92	18.92	6.18	6.18
5	14.19	14.19	5.06	4.49
10	18.92	18.92	10.67	8.99
15	18.92	18.92	8.99	6.74
20	21.62	21.62	7.87	7.87
25	18.92	17.57	6.74	6.18
30	17.57	16.22	7.30	7.30
35	20.27	20.95	6.18	5.62
40	22.30	22.3	5.62	6.74
45	21.62	21.62	8.99	7.30
50	22.30	22.30	7.87	7.30
55	19.59	19.59	9.55	8.99
60	21.62	22.97	8.43	7.30
65			7.87	10.11

of missing attribute values. Note also that there is not a big difference between certain and possible rule sets with the exception of certain rule sets and do not care conditions, where the error rate is large due to empty lower approximations for large percentage of do not care conditions.

Additionally, exact error rates are presented for fife data sets: (*breast, hepatitis, iris, lymphography* and *wine*) for missing attribute values of the type *lost*. Most surprisingly, from Tables 3–4 it is clear that for some percentage of lost values the error rate is smaller than for complete data sets (0% of lost values).

7   Conclusions

As follows from our experiments, rule sets induced from some data sets may be improved by replacing specified attribute values by missing attribute values of the type *lost*. Thus, in the process of data mining it makes sense to try to replace some portion of specified attribute values by missing attribute values of the type *lost* and check whether the error rate decreases. Replacing some portion of specified attribute values by missing attribute values of the type *lost* corresponds to hiding from the rule induction system some information. It appears that for some data sets the rule induction system, outputting new rule sets, occasionally finds better regularities, hidden in the remaining information. Additionally, the fact that error rate does not increase with replacement of larger and larger portions of specified attribute values by missing ones testifies that rough-set approach to missing attribute values is very good.

References

1. Grzymala-Busse, J.W., Grzymala-Busse, W.J.: An experimental comparison of three rough set approaches to missing attribute values. In: Peters, J.F., Skowron, A., Düntsch, I., Grzymała-Busse, J.W., Orłowska, E., Polkowski, L. (eds.) Transactions on Rough Sets VI. LNCS, vol. 4374, pp. 31–50. Springer, Heidelberg (2007)

2. Grzymala-Busse, J.W., Wang, A.Y.: Modified algorithms LEM1 and LEM2 for rule induction from data with missing attribute values. In: Proceedings of the Fifth International Workshop on Rough Sets and Soft Computing (RSSC 1997) at the Third Joint Conference on Information Sciences (JCIS 1997), pp. 69–72 (1997)

3. Stefanowski, J., Tsoukias, A.: On the extension of rough sets under incomplete information. In: Zhong, N., Skowron, A., Ohsuga, S. (eds.) RSFDGrC 1999. LNCS (LNAI), vol. 1711, pp. 73–82. Springer, Heidelberg (1999)

4. Stefanowski, J., Tsoukias, A.: Incomplete information tables and rough classification. Computational Intelligence 17, 545–566 (2001)

5. Grzymala-Busse, J.W., Hu, M.: A comparison of several approaches to missing attribute values in data mining. In: Ziarko, W.P., Yao, Y. (eds.) RSCTC 2000. LNCS (LNAI), vol. 2005, pp. 340–347. Springer, Heidelberg (2001)

6. Grzymala-Busse, J.W.: Three approaches to missing attribute values—a rough set perspective. In: Proceedings of the Workshop on Foundation of Data Mining, in conjunction with the Fourth IEEE International Conference on Data Mining, pp. 55–62 (2004)

7. Grzymala-Busse, J.W.: On the unknown attribute values in learning from examples. In: Raś, Z.W., Zemankova, M. (eds.) ISMIS 1991. LNCS, vol. 542, pp. 368–377. Springer, Heidelberg (1991)

8. Kryszkiewicz, M.: Rough set approach to incomplete information systems. In: Proceedings of the Second Annual Joint Conference on Information Sciences, pp. 194–197 (1995)

9. Kryszkiewicz, M.: Rules in incomplete information systems. Information Sciences 113, 271–292 (1999)

10. Grzymala-Busse, J.W.: Rough set strategies to data with missing attribute values. In: Workshop Notes, Foundations and New Directions of Data Mining, in conjunction with the 3-rd International Conference on Data Mining, pp. 56–63 (2003)

11. Grzymala-Busse, J.W.: Data with missing attribute values: Generalization of indiscernibility relation and rule induction. Transactions on Rough Sets 1, 78–95 (2004)

12. Grzymala-Busse, J.W.: Characteristic relations for incomplete data: A generalization of the indiscernibility relation. In: Proceedings of the Fourth International Conference on Rough Sets and Current Trends in Computing, pp. 244–253 (2004)

13. Lin, T.Y.: Topological and fuzzy rough sets. In: Slowinski, R. (ed.) Intelligent Decision Support. Handbook of Applications and Advances of the Rough Sets Theory, pp. 287–304. Kluwer Academic Publishers, Dordrecht (1992)

14. Slowinski, R., Vanderpooten, D.: A generalized definition of rough approximations based on similarity. IEEE Transactions on Knowledge and Data Engineering 12, 331–336 (2000)

15. Yao, Y.Y.: Relational interpretations of neighborhood operators and rough set approximation operators. Information Sciences 111, 239–259 (1998)

16. Grzymala-Busse, J.W., Rzasa, W.: Local and global approximations for incomplete data. In: Greco, S., Hata, Y., Hirano, S., Inuiguchi, M., Miyamoto, S., Nguyen, H.S., Słowiński, R. (eds.) RSCTC 2006. LNCS (LNAI), vol. 4259, pp. 244–253. Springer, Heidelberg (2006)

17. Grzymala-Busse, J.W., Rzasa, W.: Definability of approximations for a generalization of the indiscernibility relation. In: Proceedings of the 2007 IEEE Symposium on Foundations of Computational Intelligence (IEEE FOCI 2007), pp. 65–72 (2007)

18. Wang, G.: Extension of rough set under incomplete information systems. In: Proceedings of the IEEE International Conference on Fuzzy Systems (FUZZ_IEEE 2002), pp. 1098–1103 (2002)
19. Grzymala-Busse, J.W., Grzymala-Busse, W.J.: Improving quality of rule sets by increasing incompleteness of data sets. In: Proceedings of the Third International Conference on Software and Data Technologies, pp. 241–248 (2008)
20. Pawlak, Z.: Rough Sets. Theoretical Aspects of Reasoning about Data, Kluwer Academic Publishers, Dordrecht (1991)
21. Pawlak, Z.: Rough sets. International Journal of Computer and Information Sciences 11, 341–356 (1982)
22. Grzymala-Busse, J.W.: LERS—a system for learning from examples based on rough sets. In: Slowinski, R. (ed.) Intelligent Decision Support. Handbook of Applications and Advances of the Rough Set Theory, pp. 3–18. Kluwer Academic Publishers, Dordrecht (1992)
23. Grzymala-Busse, J.W.: A new version of the rule induction system LERS. Fundamenta Informaticae 31, 27–39 (1997)
24. Grzymala-Busse, J.W.: Knowledge acquisition under uncertainty—A rough set approach. Journal of Intelligent & Robotic Systems 1, 3–16 (1988)
25. Chan, C.C., Grzymala-Busse, J.W.: On the attribute redundancy and the learning programs ID3, PRISM, and LEM2. Technical report, Department of Computer Science, University of Kansas (1991)
26. Grzymala-Busse, J.W.: MLEM2: A new algorithm for rule induction from imperfect data. In: Proceedings of the 9th International Conference on Information Processing and Management of Uncertainty in Knowledge-Based Systems (IPMU 2002), pp. 243–250 (2002)
27. Booker, L.B., Goldberg, D.E., Holland, J.F.: Classifier systems and genetic algorithms. In: Carbonell, J.G. (ed.) Machine Learning. Paradigms and Methods, pp. 235–282. MIT Press, Boston (1990)
28. Holland, J.H., Holyoak, K.J., Nisbett, R.E.: Induction. Processes of Inference, Learning, and Discovery. MIT Press, Boston (1986)
29. Stefanowski, J.: Algorithms of Decision Rule Induction in Data Mining. Poznan University of Technology Press, Poznan (2001)

Anomaly Detection Using Behavioral Approaches

Salem Benferhat and Karim Tabia

CRIL - CNRS UMR8188, Université d'Artois
Rue Jean Souvraz SP 18 62307 Lens Cedex, France
{benferhat,tabia}@cril.univ-artois.fr
http://www.cril.univartois.fr

Abstract. Behavioral approaches, which represent normal/abnormal activities, have been widely used during last years in intrusion detection and computer security. Nevertheless, most works showed that they are ineffective for detecting novel attacks involving new behaviors. In this paper, we first study this recurring problem due on one hand to inadequate handling of anomalous and unusual audit events and on other hand to insufficient decision rules which do not meet behavioral approach objectives. We then propose to enhance the standard decision rules in order to fit behavioral approach requirements and better detect novel attacks. Experimental studies carried out on real and simulated *http* traffic show that these enhanced decision rules improve detecting most novel attacks without triggering higher false alarm rates.

Keywords: Intrusion detection, Behavioral approaches, Bayesian network classifiers, Decision trees.

1 Introduction

Intrusion detection's objective is detecting in real-time or in off-line mode any malicious activity or action compromising integrity, confidentiality or availability of computer and network resources or services [1]. Intrusion detection systems (IDSs) are either misuse-based [18] or anomaly-based [12] or a combination of both the approaches in order to exploit their mutual complementarities [19]. Behavioral approaches, which are variants of anomaly-based approaches, build profiles representing normal and abnormal behaviors and detect intrusions by comparing system activities with learnt profiles. The main advantage of these approaches lies their capacity to detect both known and novel attacks. However, there is no behavioral approach ensuring acceptable tradeoff between novel attack detection and underlying false alarm rate.

Intrusion detection can be viewed as a classification problem in order to classify audit events (network packets, Web server logs, system logs, etc.) as normal events or attacks [10]. During last years, several works used classifiers in intrusion detection [8],[16],[20] and achieved acceptable detection rates on well-known benchmarks such as KDD'99 [10], Darpa'99 [11]. The recurring problem with the majority of classifiers is their high false negative rates mostly caused by the incapacity to correctly classify novel attacks [5][10]. For instance, in [4][2], decision trees and variants of Bayes classifiers are used to classify network connections and concluded that their main problem lies in their failure to detect novel attacks which they classify normal connections.

J. Cordeiro et al. (Eds.): ICSOFT 2008, CCIS 47, pp. 217–230, 2009.

In this paper, we first analyze and explain the problem of high false negative rates and classifiers incapacity to correctly classify new malicious behaviors. We illustrate our work with Bayesian network classifiers and decision trees. We first focus on how new behaviors affect and manifest through a given feature set. Then we explain why standard decision rules fail in detecting these new events. More precisely, we consider on one hand problems related to handling unusual and new behaviors and on other hand problems due to insufficient decision rules which do not meet anomaly detection requirements. After that, we propose to enhance standard decision rules in order to improve detecting novel attacks involving abnormal behaviors. Experimental studies on real and simulated $http$ traffic are carried out to evaluate the effectiveness of the new decision rules in detecting new intrusive behaviors. Two variants of Bayesian classifiers and C4.5 decision tree using the enhanced decision rules are trained on real normal $http$ traffic and several Web attacks. Then we evaluated these classifiers on known and novel attacks as well new normal behaviors.

The rest of this paper is organized as follows: Section 2 presents Behavioral approaches focusing on Bayesian classifiers and decision trees and their inadequacy for detecting novel attacks. Section 3 proposes enhancements to the standard Bayesian classification in order to improve detecting novel attacks while Section 4 proposes other enhancements for decision trees. Experimental studies on $http$ traffic are presented in section 5. Section 6 concludes this paper.

2   Behavioral Approaches for Intrusion Detection

Behavioral approaches build models or profiles representing normal and abnormal activities and detect intrusions by computing deviations of current system activities form reference profiles. Behavioral approaches are like anomaly detection ones except that profiles are defined for both normal and abnormal behaviors while in anomaly approaches, only normal behaviors are profiled[1]. For instance, every significant deviation from normal behavior profiles may be interpreted as an intrusion since it represents an anomalous behavior. The main advantage of anomaly approaches lies in their potential capacity to detect both new and unknown (previously unseen) attacks as well as known ones. This is particularly critical since new attacks appear every day and it often takes several days between the apparition of a new attack and updating signature data bases or fixing/correcting the exploit.

Behavioral approaches can be viewed as classifiers which are mapping functions from a discrete or continuous feature space (observed variables $A_0 = a_0$, $A_1 = a_1$, .., $A_n = a_n$) to a discrete set of class labels $C=\{c_1, c_2,..,c_m\}$. Once a classifier is built on labeled training data, it can classify any new instance. The goal of classification is to maximize the generalization ability to correctly classify unseen instances. Decision trees[13] and Bayesian classifiers [6] are well-known classification algorithms. Note that when intrusion detection is modeled as a classification problem, each instance to classify represents an audit event (network packet, connection, application log record, etc.).

In order to analyze standard classification rules incapacity to detect novel attacks, we first focus on how novel attacks involving new behaviors affect feature sets which provide input data to be analyzed.

2.1 Why Standard Classification Rules Are Ineffective for Detecting Novel Attacks

Within anomaly detection approaches, the detection of novel attacks has several negative side effects which concern triggering very high false alarm rates. This drawback seriously limits there use in real applications. In fact, configuring anomaly-based IDSs to acceptable false alarm rates cause their failure in detecting most malicious behaviors.

How Novel Attacks Affect Feature Sets. The following are different possibilities about how novel events affect and manifest through feature sets:

1. *New Value(s) in a Feature(s).* A never seen[1] value is anomalous and it is due in most cases to a malicious event. For example, Web server response codes are from a fixed set of predefined values (ex. 200, 404, 500...). If a new response code or any other response is observed, then this constitutes an anomalous event. For instance, successful shell code attacks cause server response without a common code. Similarly, a new network service using a new and uncommon port number is probably intrusive since most back-door attacks communicate through uncommon ports while common services are associated with common port numbers.

2. *New Combination of Known Values.* In normal audit events, there are correlations and relationships between features. Then an anomalous event can be in the form of a never seen combination of normal values. For example, in some *http* requests, numerical values are often provided as parameters. The same values which are correctly handled by a given program, can in other contexts cause value misinterpretations and result in anomalous behaviors. Another example from network intrusion field is when a network service like *http* uses uncommon transport protocol like UDP. Both *http* and UDP are common protocols in network traffic. However, *http* service uses TCP protocol at the transport layer and never UDP, then *http* using UDP is anomalous event.

3. *New Sequence of Events.* There are several normal audit events which show sequence patterns. For example, in on-line marketing applications, users are first authenticated using *https* protocol for confidential data transfers. Then a user session beginning without *https* authentication is probably intrusive since the application control flow has not been followed. Such intrusive evidence can be caught by history features summarizing past events or by using appropriate mining anomaly sequence patterns algorithms.

4. *No Anomalous Evidence.* In this case, new anomalous events do not result in any unseen evidence. The underlying problem here is related to feature extraction and selection since not enough data is used for catching the anomalous evidence.

In principle, the three first possibilities can be detected since abnormal evidence had appeared in the feature set. However, anomalous audit event of fourth case can not be detected for lack of any anomalous evidence in the audit event.

---

[1] By never seen value we mean new value in case of nominal features or very deviating value in case of numerical features.

Why Novel Attacks Cause False Negatives. Novel attacks often involve new behaviors. However, in spite of these anomalousness evidence in the feature set, most classifiers flag novel attacks as normal events. This failure is mainly due to the following problems:

1. *Inadequate Handling of New and Unusual Events.* New and unusual values or value combinations are often involved by novel attacks. However, most classifiers handle such evidence inadequately regarding anomaly detection objectives. For instance, in Bayesian classifiers [6], new values cause zero probabilities which most implementations replace with extremely small values and rely on remaining features in order to classify the instance in hand. Decision trees [13], which are very efficient classifiers, often use few features in order to classify audit events. Therefore, if the anomalous evidence (abnormal values or value combinations) appears in a feature which is not used, then this evidence is not used. Other techniques suffer from other problems such as incapacity to handle categorical features which are common in intrusion detection [17].
2. *Insufficient Decision Rules.* The objective of standard classification rules is to maximize classifying previously unseen instances relying on known (training) behaviors. However, unseen behaviors which should be flagged abnormal according to anomaly approach, are associated with known behavior classes. For instance, when decision trees classify instances, new behaviors such as new values are assigned to the majority class at the test where the new value is encountered. As for Bayesian classifiers, they rely only on likelihood and prior probabilities to ensure classification. This strongly penalize detection of new and unusual behaviors in favor of frequent and common behaviors. Given that normal data often represent major part of training data sets [10],[11], standard classification rules fail in detecting novel attacks involving new behaviors and flag them in most cases normal events.

In order to overcome standard behavioral approaches' drawbacks, we propose enhancing them in order to fit behavioral approach requirements. In the following, we will only focus on Bayesian networks and decision trees as examples of behavioral approaches. This choice is motivated by the fact that they are among most effective techniques and because of their capacity to handle both numeric and categorical features which are common in intrusion detection [17].

3   Enhancing Bayesian Classification for Anomaly Detection

Bayesian classification is a particular kind of Bayesian inference [6]. Classification is ensured by computing the greatest a posteriori probability of the class variable given an attribute vector. Namely, having an attribute vector A (observed variables $A_0 = a_0$, $A_1 = a_1$, .., $A_n = a_n$), it is required to find the most plausible class value c_k ($c_k \in$ C=$\{c_1, c_2,..,c_m\}$) for this observation. The class c_k associated to A is the class with the most a posteriori probability $p(c_k/A)$. Then Bayesian classification rule can be written as follows:

$$Class = argmax_{c_k \in C}(p(c_i/A)) \tag{1}$$

Term $p(c_i/A)$ denotes the posterior probability of having class c_i given the evidence A. This probability is computed using Bayes rule as follows:

$$p(c_i/A) = \frac{p(A/c_i) * p(c_i)}{p(A)} \tag{2}$$

In practice, the denominator of Equation 2 is ignored because it does not depend on the different classes. Equation 2 means that posterior probability is proportional to likelihood and prior probabilities while evidence probability is just a normalizing constant. Naive Bayes classifier assumes that features are independent in the class variable context. This assumption leads to the following formula

$$p(c_i/A) = \frac{p(a_1/c_i) * p(a_2/c_i)..p(a_n/c_i) * p(c_i)}{p(A)} \tag{3}$$

In the other Bayesian classifiers such as TAN (Tree Augmented Naive Bayes), BAN (Augmented Naive Bayes) and GBN (General Bayes Network) [6], Equation 2 takes into account feature dependencies in computing conditional probabilities as it is denoted in Equation 4.

$$p(c_i/A) = \frac{p(a_1/Pa(a_1)) * .. * p(a_n/Pa(a_n)) * p(c_i)}{p(A)} \tag{4}$$

Note that terms $Pa(a_i)$ in Equation 4 denote parents of feature a_i.

Bayesian classifiers have been widely used in intrusion detection. For instance, in [20], naive Bayes classifier is used to detect malicious audit events while in [8], authors use Bayesian classification in order to improve the aggregation of different anomaly detection model outputs.

Bayesian classification lies on posterior probabilities given the evidence to classify (according to Equations 1 and 2). The normality associated with audit event E (observed variables $E_0 = e_0$, $E_1 = e_1$, .., $E_n = e_n$) can be measured by posterior probability $p(Normal/E)$. This measure is proportional to the likelihood of E in $Normal$ class and prior probability of $Normal$ class.

In practice, normality can not be directly inferred from probability $p(Normal/E)$ because this probability is biased. For instance, major Bayesian classifier implementations ignore denominator of Equation 2 while zero probability and floating point underflow problems are handled heuristically. Assume for instance that a never seen value had appeared in a nominal feature e_i. Then according to Equation 2, the probability $p(e_i/c_k)$ equals zero over all classes c_k. In practice, it is an extremely small value that is assigned to this probability. The strategy of assigning non zero probabilities in case of new values is to use remaining features and prior probabilities in order to classify the instance in hand. The other problem consists in floating point underflow which is caused by multiplying several small probabilities each varying between 0 and 1. This case is often handled by fixing a lower limit when multiplying probabilities.

3.1 Using Zero Probabilities as Abnormal Evidence

Anomalous audit events can affect the feature set either by new values, new value combinations or new audit event sequences. Then classifying anomalous events strongly depends on how zero probability and floating point underflow problems are dealt with. However, since a zero probability is due to new (hence anomalous) value, then this is anomalousness evidence. The underlying interpretation is that instance to classify involves a never seen evidence. Then anomaly approach should flag this audit event anomalous. Similarly, an extremely small a posteriori probability can be interpreted as a very unusual event, hence anomalous. Then, standard Bayesian classification rule can accordingly be enhanced in the following way:

- If there is a feature e_i where probability $p(e_i/c_k)$ equals zero over all training classes, then this is a new value (never seen in training data). Enhanced Bayesian classification rule can be formulated as follows:

Rule 1

$$\text{If } \exists\ e_i,\ \forall k, p(e_i/c_k) = 0 \text{ then } Class = New$$
$$\text{else } Class = argmax_{c_k \in C}(p(c_k/E))$$

- New intrusive behaviors can be in the form of unseen combination of seen values. In this case, feature dependencies must be used in order to reveal such anomalousness. Since new value combinations will cause zero conditional probabilities, then this anomalous evidence can be formulated as follows:

Rule 2

$$\text{If } \exists\ e_i,\ p(e_i/Pa(e_i)) = 0 \text{ then } Class = New$$
$$\text{else } Class = argmax_{c_k \in C}(p(c_k/E))$$

Note that when building Bayesian classifiers, structure learning algorithms extract feature dependencies from training data. Then there may be unseen value combinations that can not be detected if the corresponding dependencies are not extracted during structure learning phase.

3.2   Using Likelihood of Rare Attacks as Abnormal Evidence

When training classifiers, some attacks have often very small frequencies in training data sets. The problem with such prior probabilities is to strongly penalize the corresponding attacks likelihood. This problem was pointed out in [3] where authors proposed simple duplication of weak classes in order to enhance their prior probabilities. An alternative solution is to exploit likelihood of audit events as if training classes ($Normal, Attack_1,.., Attack_n$) were equiprobable. Assume for instance intrusive audit event E is likely to be an attack (for example, likelihood $p(E/Attack_j)$ is the most important). Because of the negligible prior probability of $Attack_j$, posterior probability $p(Attack_j/E)$ will be extremely small while $p(Normal/E)$ can be significant since $Normal$ class prior probability is important. Then we can rely on likelihood in order to detect attacks with small frequencies:

Rule 3

$$\textbf{If } \exists \; Attack_j, \;\; \forall k, p(E/Attack_j) >= p(E/c_k) \textbf{ and}$$
$$p(Normal/E) > P(Attack_j/E) \textbf{ and } p(Attack_j) < \epsilon$$
$$\textbf{then} \quad Class = Attack_j$$
$$\textbf{else} \;\; Class = argmax_{c_k \in C}(p(c_k/E))$$

Rule 3 is provided in order to help detecting anomalous events with best likelihood in attacks having extremely small prior probabilities ($p(Attack_j) < \epsilon$). It will be applied only if the proportion of instances of $Attack_j$ in training data is less than threshold ϵ fixed by the expert.

Note that standard Bayesian classification rule (see Equation 1) is applied only if Rules 1, 2 and 3 can not be applied. As for the priority for applying these rules, we must begin by zero probability rules (Rules 1 and 2) then likelihood rule (Rule 3).

4 Enhancing Decision Trees for Anomaly Detection

4.1 Decision Tree Classifiers

Decision trees [13][14] are among the most efficient classification algorithms. Three node types compose a decision tree: (1) A unique root node, (2) a set of decision nodes (internal nodes) and (3) a set of terminal nodes called leaves.

Except the leaves, each node in a decision tree represents a test that splits items of training set at that node into more homogenous partitions. For every possible test modality, there will be a child partition created especially to handle items satisfying that test. Leaves contain items that most belong to the same class. Learning a decision tree consists of recursively partitioning the training data (pre-classified data) into smaller partitions in such a way to improve homogeneity (proportion of items belonging to the same class) of child partitions. The choice of splitting criteria at a given node is concerned with selecting the test attribute and, if necessary, its value that most discriminates between items at the considered node. All the splitting criteria used in practice by decision tree induction algorithms are variants of impurity or entropy measures that evaluate the efficiency of a test by comparing the entropy at the node to split and its potential child partitions. In order to classify a new instance, we start by the root of the decision tree. Then we test the attribute specified by this node. The result of this test allows moving down the tree through the branch relative to the attribute value of the given instance. This process is repeated until a leaf is encountered. The instance is then classified in the same class as the one characterizing the reached leaf.

In our experimentations, we used C4.5 [14] which is an efficient decision tree induction algorithm. This technique builds n-ary decision trees using $GainRatio$ as splitting criteria and associates the majority class to a leaf.

4.2 Decision Tree Adaptations for Anomaly Detection

Standard decision tree algorithms aim at minimizing both misclassification rate and tree size [14]. The minimum description length principle (MDL) [21] is used for model selection in many machine learning algorithms such as decision trees, Bayesian network structure learning, etc. This principle, which was formalized by Rissanen [21], states that the "best" model is the shortest one. Note that MDL principle is a formalization of the Occam's razor which prefers among several hypothesis the simplest one. In machine learning algorithms, this results in preferring the shortest and most concise model (in terms of model size). In decision tree algorithms, MDL is used for model selection and leads to preferring shorter trees (often in terms of node number). We can still build decision trees that minimize misclassification rate while partially (or not at all) minimizing tree size. The motivation of not minimizing tree size is to build trees with more test nodes in such a way that more tests will be performed when classifying items. In fact, the greater the number of tests (test nodes) performed during classification procedure, the better will be the probability of examining features where anomalous behavior evidence occurs.

Compatible Decision Trees. We use term "Compatible decision trees" to denote decision trees built using same algorithm on same training data (same feature space and same training data distribution) but with variants of attribute selection measures or using different stopping/pruning options. For example, attribute selection measure used in C4.5 algorithm selects the feature with greatest gain ratio at each decision node. Instead, one can select the feature with second best gain ratio (or with any other gain ratio ranking). Figure 1 shows two compatible decision trees built on exactly same training set. In trees of Figure 1, elliptic nodes represent test nodes while rectangular ones represent leaves. Note that the first tree (Tree 1) is built using the standard C4.5 algorithm while the second tree (Tree 2) is built using C4.5 algorithm with attribute selection measure modified such that it selects the attribute having the second best gain ratio. The second tree still minimizes misclassification rate while on purpose it does not minimize tree size in order to increase the number of tests required to classify an instance. Indeed, both the trees correctly classify all training instances but the size of the first tree equals 8 nodes (3 test nodes and 5 leaves) while the second tree includes 13 nodes (5 test nodes and 8 leaves). Note also that in the first tree, there are instances which are classified by performing only one test (on feature $outlook$) while at most two tests are needed to classify any instance. However, in the second tree, two tests are at least needed to classify any instance and three tests are at most needed to perform classification. The reason for such a difference lies in the feature selection measure. In the example of Figure 1, the first tree complies with MDL principle while the second tree relaxes MDL principle but still minimizes misclassification rate. Note that longer trees can be built by selecting features with weaker gain ratio.

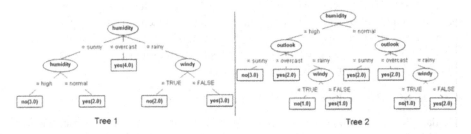

Fig. 1. Example of compatible decision trees

When used as a behavioral approach, the advantage of using relaxed MDL decision trees is that longer trees imply more features to be tested when classifying instances. Hence, if novel/abnormal behavior's evidence occurs through a given feature, it is more likely that this feature will be tested in longer trees than in short standard trees.

Relaxing MDL Principle. The strategy of relaxing MDL while building decision trees is to increase the number of tests to be performed when classifying audit events. However, this will result in trees of great size penalizing CPU/Memory usage and cause tree overfitting. Then a tradeoff between tree size and tree overfitting is needed and can be fixed experimentally.

1 - Relaxing Tree Pruning/Stopping Criteria. Stopping criteria and pruning often result in gathering within same leaves instances labeled *normal* and *attacks*. And because of associating majority class to leaves, these latter are in most cases labeled *normal* since the number of normal instances are often larger than attack ones in training sets. This is for instance one among the reasons causing high false negative rates relative to decision trees. Then, instead of misclassifying these attacks, the growth of the tree should be continued until obtaining leaves containing either normal items or attacks (even if attacks belonging to different classes) but never normal items with attacks. Indeed, only leaves containing either only normal items or only attacks guarantee avoiding false negatives and false alarms. Similarly, pruning often results in engendering leaves involving normal and attack instances. Consequently, it is better to cancel pruning phase in order to avoid generating leaves melting normal and attack instances. Such enhancements and details may seem insignificant but regarding the huge numbers of audit events (network packets, system log records, etc.) that are to be daily classified, misclassifications due to these inadequate stopping criteria and pruning options can cause unacceptable false alarm and false negative rates. For instance, in order to guarantee a 100 false alarms rate per day, it is required to ensure a misclassification rate less than 0.01% in a system dealing with one million audit events per day (which is very common in nowadays network intrusion detection systems).

2 - Relaxing Feature Selection Measures. The feature selection measure used by C4.5 algorithm selects the feature having best gain ratio. Let us assume that this feature selection measure selects another feature with less gain ratio. This will likely result in smaller gain ratio and will require more tests in order to obtain leaves[2]. Then, instead of using the feature with best gain ratio, we can build compatible decision trees using a feature selection measure selecting for instance the feature with average gain ratio. Let D be the set of m labeled training instances and $A_1, A_2,..,A_n$ be the features describing instances. Let also $GainRatio(T, A_i)$ be the function computing the gain ratio achieved if training instances involved by the current set of items T are split using A_i. Attributes can then be ranked according to their gain ratio and we can select any attribute from the one having best gain ratio to the one with the worse gain ratio. Namely,

$$Feature = RankSelectionMeasure(GainRatio(T, A_1), .., GainRatio(T, A_n), Rank) \quad (5)$$

---

[2] A node is declared a leaf if it is pure or when it satisfies the stopping criteria

Feature selection measure of Equation 5 computes and ranks the gain ratios achieved by features $A_1,..,A_n$ on the set of current training items T and returns the feature having the specified ranking $Rank$ ($1 \leq Rank \leq n$). Selecting features with best gain ratios will induce very short trees while selecting attributes with weak gain ratios will result in extremely long trees which heavily penalizes both learning and classification procedures in terms of CPU/memory usage. It is clear that a tradeoff between tree size and generalization/overfitting aspects must be fulfilled taking into account intrusion detection specificities. As it will be shown in experimental studies, selecting attributes with average gain ratios ensures a good tradeoff between tree size and misclassification rate.

5 Experimental Studies

In this section, we provide experimental studies of our enhanced Bayesian classification rule on $http$ traffic including normal real data and several Web attacks. Before giving further details, we first present training and testing data then we report evaluation results of naive Bayes and Tree Augmented naive Bayes (TAN), the two Bayesian classifiers we used in the following experimentations.

5.1 Training and Testing Data Sets

Our experimental studies are carried out on Web-based attack detection problem which represents major part of nowadays cyber-attacks. In [22], we proposed a set of detection models including basic features of $http$ connections as well as derived features summarizing past $http$ connections and providing useful information for revealing suspicious behaviors involving several $http$ connections. The detection model features are grouped into four categories:

1. **Request General Features.** They are features that provide general information on $http$ requests. Examples of such features are request method, request length, etc.
2. **Request Content Features.** These features search for particularly suspicious patterns in $http$ requests. The number of non printable/metacharacters, number of directory traversal patterns, etc. are examples of features describing request content.
3. **Response Features.** Response features are computed by analyzing the $http$ response to a given request. Examples of these features are response code, response time, etc.
4. **Request History Features.** They are statistics about past connections given that several Web attacks such as flooding, brute-force, Web vulnerability scans perform through several repetitive connections. Examples of such features are the number/rate of connections issued by same source host and requesting same/different URIs.

Note that in order to label the preprocessed $http$ traffic (as normal or attack), we analyzed this data using Snort[18] IDS as well as manual analysis. As for other attacks, we simulated most of the attacks involved in [7] which is to our knowledge the most extensive and uptodate open Web-attack data set. In addition, we played vulnerability

Table 1. Training/testing data set distribution

	Training data		Testing data	
Class	Number	%	Number	%
Normal	55342	55.87%	61378	88.88 %
Vulnerability scan	31152	31.45%	4456	6.45 %
Buffer overflow	9	0.009%	15	0.02%
Input validation	44	0.044%	4	0.01 %
Value misinterpretation	2	0.002%	1	0.00%
Poor management	3	0.003%	0	0.00%
URL decoding error	3	0.003%	0	0.00%
Flooding	12488	12.61%	3159	4.57 %
Cross Site Scripting	**0**	**0.00%**	**6**	**0.0001 %**
SQL injection	**0**	**0.00%**	**14**	**0.001 %**
Command injection	**0**	**0.00%**	**9**	**0.001 %**
Total	**99043**	**100%**	**69061**	**100%**

scanning sessions using w3af[15]. Note that attack simulations are achieved on a simulation network using the same platform (same Web server software and same operating system) and same Web site content.

Attacks of Table 1 are categorized according to the vulnerability category involved in each attack. Regarding attack effects, attacks of Table 1 include DoS attacks, Scans, information leak, unauthorized and remote access [7]. In order to evaluate the generalization capacities and the ability to detect new attacks, we build a testing data set including real normal *http* connections as well as known attacks, known attack variations and novel ones (attacks in bold in Table 1).

Note that new attacks included in testing data either involve new feature values or anomalous value combinations:

- *Attacks Causing new Values.* New attacks involving new values are SQL injection, XSS and command injection attacks. These attacks are characterized by never seen values in training data.
- *Attacks Causing New Value Combinations.* New attacks involving new value combinations are buffer overflow attacks including shell code and shell commands and vulnerability scans searching particularly for SQL injection points. In training data, vulnerability scans search for vulnerabilities other than SQL injection points. Similarly, in training data, there are shell command injections and buffer-overflows without shell codes or shell commands.

Note that SQL injection attacks include two insufficient authentication attacks performing through SQL injection. These attacks cause simultaneously anomalous new values and value combinations.

5.2 Experiments on Standard/Enhanced Bayesian Classification Rule

Table 2 compares results of standard then enhanced naive Bayes and TAN classifiers built on training data and evaluated on testing one.

Note that enhanced classification rule evaluated in Table 2 uses normality/abnormality duality and zero probabilities (see Rules 1, 2 and 3).

Table 2. Evaluation of naive Bayes and TAN classifiers using standard/enhanced Bayesian classification rules

	Standard Bayesian rule		Enhanced Bayesian rule	
	Naive Bayes	TAN	Naive Bayes	TAN
Normal	98.2%	99.9%	91.7%	97.8%
Vulnerability scan	15.8%	44.1%	100%	100%
Buffer overflow	6.7%	20.2%	80%	100%
Input validation	75.0%	100%	100%	100%
Value misinterpretation	100%	0.00%	100%	100%
Flooding	100%	100%	100%	100%
Cross Site Scripting	**0.00%**	**0.00 %**	**100%**	**100%**
SQL injection	**0.00%**	**0.00%**	**100 %**	**100%**
Command injection	**0.00%**	**0.00 %**	**100 %**	**100%**
Total PCC	**92.87%**	**96.24%**	**96.45%**	**98.07%**

– *Experiments on Standard Bayesian Classification Rule.* At first sight, both classifiers achieve good detection rates regarding their PCCs (Percent of Correct Classification) but they are ineffective in detecting novel attacks (attacks in bold in Table 2). Confusion matrixes relative to this experimentation show that naive Bayes and TAN classifiers misclassified all new attacks and predicted them $Normal$. However, results of Table 2 show that TAN classifier performs better than naive Bayes since it represents some feature dependencies. Furthermore, testing attacks causing new value combinations of seen anomalous values (involved separately in different training attacks) cause false negatives. For instance, testing vulnerability scans are not well detected since they involve new value combinations.

– *Experiments on Enhanced Bayesian Classification Rule.* Naive Bayes and TAN classifiers using the enhanced rule perform significantly better than with standard rule. More particularly, both the classifiers succeeded in detecting both novel and known attacks. Unlike naive Bayes, enhanced TAN classifier improves detection rates without triggering higher false alarm rate (see correct classification rate of $Normal$ class in Table 2. Furthermore, TAN classifier correctly detects and identifies all known and novel attacks.

Results of Table 2 show that significant improvements can be achieved in detecting novel attacks by enhancing standard classification rules in order to meet behavioral approach requirements.

Because of page number limitation, we were not capable to provide our experimental studies on enhancements for decision trees which achieved significant improvements in detecting novel and rare attacks.

6   Conclusions

The main objective of this paper is to overcome one of the main limitations of behavioral approaches. We proposed to enhance standard decision rules in order to effectively detect both known and novel attacks. We illustrated our enhancements on Bayesian and decision tree classifiers in order to improve detecting novel attacks involving abnormal behaviors. More precisely, we have proposed for Bayesian classifiers new rules exploiting zero probabilities caused by anomalous evidence occurrence and likelihood of attacks having extremely small prior frequencies. As for decision trees, we proposed to relax MDL principle and avoid gathering within a same leaf normal and attack instances. Experiments on *http* traffic show the significant improvements achieved by the enhanced decision rule in comparison with the standard one. Future work will address handling incomplete and uncertain information relative to network traffic audit events.

Acknowledgements. This work is supported by a French national project entitled DADDi (Dependable Anomaly Detection with Diagnosis).

References

1. Axelsson, S.: Intrusion detection systems: A survey and taxonomy. Technical Report 99-15, Chalmers Univ. (2000)
2. Barbará, D., Wu, N., Jajodia, S.: Detecting novel network intrusions using bayes estimators. In: Proceedings of the First SIAM Conference on Data Mining (2001)
3. Ben-Amor, N., Benferhat, S., Elouedi, Z.: Naive bayesian networks in intrusion detection systems. In: ACM, Cavtat-Dubrovnik, Croatia (2003)
4. Benferhat, S., Tabia, K.: On the combination of naive bayes and decision trees for intrusion detection. In: CIMCA/IAWTIC, pp. 211–216 (2005)
5. Elkan, C.: Results of the kdd 1999 classifier learning. SIGKDD Explorations 1(2), 63–64 (2000)
6. Friedman, N., Geiger, D., Goldszmidt, M.: Bayesian network classifiers. Machine Learning 29(2-3), 131–163 (1997)
7. Ingham, K.L., Inoue, H.: Comparing anomaly detection techniques for http. In: Recent Advances in Intrusion Detection, pp. 42–62 (2007)
8. Kruegel, C., Mutz, D., Robertson, W., Valeur, F.: Bayesian event classification for intrusion detection. In: 19th Annual Computer Security Applications Conference, Las Vegas, Nevada (2003)
9. Kumar, S., Spafford, E.H.: An application of pattern matching in intrusion detection. Tech. Rep. CSD–TR–94–013, Department of Computer Sciences, Purdue University, West Lafayette (1994)
10. Lee, W.: A data mining framework for constructing features and models for intrusion detection systems. PhD thesis, New York, NY, USA (1999)
11. Lippmann, R., Haines, J.W., Fried, D.J., Korba, J., Das, K.: The 1999 darpa off-line intrusion detection evaluation. Comput. Networks 34(4), 579–595 (2000)
12. Neumann, P.G., Porras, P.A.: Experience with EMERALD to date, pp. 73–80 (1999)
13. Quinlan, J.R.: Induction of decision trees. Mach. Learn. 1(1) (1986)
14. Ross Quinlan, J.: C4.5: programs for machine learning. Morgan Kaufmann Publishers Inc., San Francisco (1993)
15. Riancho, A.: w3af - web application attack and audit framework (2007)
16. Sebyala, A.A., Olukemi, T., Sacks, L.: Active platform security through intrusion detection using naive bayesian network for anomaly detection. In: Proceedings of the London Communications Symposium 2002 (2002)
17. Shyu, M.-L., Sarinnapakorn, K., Kuruppu-Appuhamilage, I., Chen, S.-C., Chang, L., Goldring, T.: Handling nominal features in anomaly intrusion detection problems. In: RIDE, pp. 55–62. IEEE Computer Society, Los Alamitos (2005)
18. Snort Snort: The open source network intrusion detection system (2002), http://www.snort.org

19. Tombini, E., Debar, H., Me, L., Ducasse, M.: A serial combination of anomaly and misuse idses applied to http traffic. In: ACSAC 2004: Proceedings of the 20th Annual Computer Security Applications Conference (ACSAC 2004), pp. 428–437. IEEE Computer Society, Washington (2004)
20. Valdes, A., Skinner, K.: Adaptive, model-based monitoring for cyber attack detection. In: Recent Advances in Intrusion Detection, pp. 80–92 (2000)
21. Rissanen, J.: Modelling by the shortest data description. Automatica 14, 465–471 (1978)
22. Benferhat, S., Tabia, K.: Classification features for detecting server-side and client-side web attacks. In: SEC 2008: 23rd International Security Conference, Milan, Italy (2008)

Author Index